Praise for
Explore

"*Explore/Create* is a chronicle of things the future may hold. F passion for space. While Richa t aboard the ISS, I am building a colonization fleet of ships to take all of us to Mars. Perhaps one day our kids will create and play games on a new world!"

—Elon Musk, CEO, SpaceX and Tesla Motors

"Richard Garriott has been inspiring me to explore and create since I was a teenager. I never would have written *Ready Player One* if I hadn't grown up playing the games he created as Lord British. This is the book I have been waiting to read for a long, long time."

—Ernest Cline, author of *Ready Player One*

"Fascinating. . . . Demystifies an industry and a man so instrumental in it. . . . A rich compendium of stories from an extraordinary life."

—Eurogamer.net

"Richard and his partners at Zero-G took me on my first flight into weightlessness, he carried the book my daughter and I co-wrote—*George's Secret Key to the Universe*—to the International Space Station and back, and soon I hope to ride to space myself aboard a Virgin Galactic vehicle, which grew out of the X-Prize that he cofounded. I applaud his spirit and have enjoyed the fruits of his labor."

—Stephen Hawking, author of *A Brief History of Time*

"Lively and entertaining. . . . [Garriott] embeds puzzles and games in the book for those who want to take it beyond the simple experience of reading. . . . Those looking for glimpses into an adventurous life should be pleased."

—*Kirkus Reviews*

"Gaming takes the virtual world in our brains and makes something real out of it. I was so taken by Richard Garriott's game *Ultima* that I was nearly speechless when I met him at a pizza joint in Chicago. Now, finally, Richard has shared his compelling story with the rest of us in *Explore/Create*."

—Steve Wozniak, cofounder of Apple

"Garriott's enthusiasm for his passions is evident and . . . even inspirational."

—*Booklist*

"This book will take you on an insider's journey into Richard's fascinating mind, from creating virtual worlds and characters to exploring the cosmos. Buckle up!"

—Anousheh Ansari, first Muslim woman in space

"Telling. . . . Learn [Garriott's] secret to staying creative and gain valuable insight into his crazy-interesting history." —Brit + Co

"The day Richard Garriott returned from space, I met him with a bouquet of flowers and a personal painting. Richard is a great example of the second generation of space explorers. He has followed his father not just into space, but as a champion of international cooperation in discovering the secrets of the universe and using these advancements for the good of humanity."

—Alexei Leonov, Russian astronaut and first person to do a spacewalk

"An entertaining book that describes Garriott's twin passions to explore (not just space but also the deep sea and Antarctica, among other places) and create." —The Space Review

Explore/Create

Explore/Create

My Life in Pursuit of New Frontiers, Hidden Worlds, and the Creative Spark

Richard Garriott

with David Fisher

WILLIAM MORROW
An Imprint of HarperCollins*Publishers*

Space helmet image © mhatzapa/Shutterstock, Inc. Light bulb image © Dooder/Shutterstock, Inc. Sheet of binary codes © iunewind/Shutterstock, Inc. Close-up of Milky Way © MaIII Themd/Shutterstock, Inc. Astronaut in space © fStop Images GmbH/Shutterstock, Inc. Unless otherwise noted, all photographs are courtesy of the author.

HarperCollins books may be purchased for educational, business, or sales promotional use. For information, please email the Special Markets Department at SPsales@harpercollins.com.

A hardcover edition of this book was published in 2017 by William Morrow, an imprint of HarperCollins Publishers.

FIRST WILLIAM MORROW PAPERBACK EDITION PUBLISHED 2018.

Designed by William Ruoto

Library of Congress Cataloging-in-Publication Data has been applied for.

ISBN 978-0-06-228666-6

18 19 20 21 22 LSC 10 9 8 7 6 5 4 3 2 1

It's easy to believe that I have simply been very lucky. But I also believe the adage that luck is the intersection of opportunity and preparation.
This book is dedicated to all those who helped me with either on this journey.

This book is also dedicated to my dear wife, Laetitia, and to our children, Kinga Shuilong and Ronin Phi.
Sharing adventures as a family brings me more joy than I had imagined.
A great man once said, "Oh the places you'll go!" I can't wait to see where.
And Kinga and Ronin—I look forward to listening to YOUR tales and visions for the future . . .

Contents

Explore/Create

Introduction

Explore and *Create*. These words are inextricable from each other in my life. They feed each other. My dual-sided business card has my exploration street cred listed on one side and my creation credentials on the other.

We are fortunate to live in such a remarkable era for exploration and creation, and I often reflect on my own luck. I was lucky to be born at the dawn of personal computers. I was lucky to be the son of a scientist and explorer—my father was a NASA astronaut—while my mother is an extraordinarily creative artist. I was lucky to have a childhood filled with exploration, a celebration of the reality in which we live. I was lucky to learn early on that a deep understanding of the world around you makes you its master.

So admittedly I've had my fair share of lucky breaks. But I do buy into the adage that luck is the intersection of preparation and opportunity. Opportunities parade past all of us all the time. The key is that you must be paying attention to see them, you must be willing to take risks, you must expose yourself to the possibility of massive failure—and you must believe in what you are doing so much that you do it anyway.

This attitude has enabled me to help create and build two world-impacting industries: computer games and commercial spaceflight. On the surface, these two areas would seem to have nothing to do with one another. But both are about exploring new frontiers, creatively envisioning ways to go places no one has before. And a key element of this creative vision is interactive storytelling.

Interactive stories are where creativity and exploration meet. It's a new art form, one I've had the pleasure of helping shape by writing interactive games that go far beyond fighting monsters, games that can probe the depth of human experience as surely as a novel or painting or other "traditional" work of art. In an interactive environment, players are free to explore, but they can explore only what the developers have placed there for them, which compels the developer to consider all possible explorations of the reality they are crafting. On my own, I created some of the first virtual worlds. I did it with little to go on, largely through trial and error. Like a good artist in any other medium, I became a polymath. I studied subjects from philosophy and religious history to architecture, languages, physics, and fashion. Because creating an interactive world that is engaging and satisfying requires a knowledge of all of its many facets.

Games have become much more than pleasant diversions. They have a huge opportunity to become *the* media form of the twenty-first century. Because games, especially role-playing games, can teach players at a deep level: invite them to examine their values and morals, not just in text or images but in realistic cause–effect scenarios that transcend linear narrative. Much as children "role-play" to learn about the world around them, through interactive games adults can capture that same sense of wonder.

That wonder has driven my career, and I hope to impart it to you here. How you explore this book is up to you. You will see that each chapter fits into one of the dual narratives of my life, either "Explore" or "Create." You can, of course, read this book straight through, but if you prefer to jump around, to follow your own passions and interests, I encourage you to do so. I hope by reading . . . no, exploring . . . the pages that follow, and taking on the challenges I offer within them, thinking about what *you* would do in many of these same situations, you will feel as though you have embarked on your own adventure.

Carpe futurum!

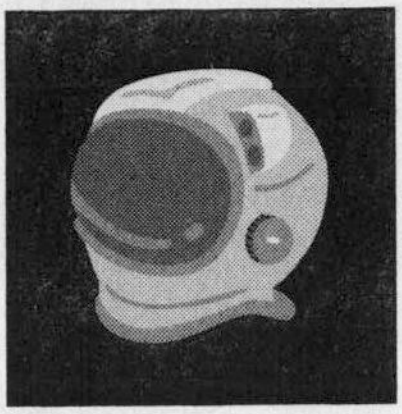

1

EXPLORE

Titanic Mistakes

We were sitting inside a twenty-ton submarine 12,600 feet beneath the surface of the North Atlantic, trapped under the crumbling stern of RMS *Titanic.* All around us a blinding silt storm raged. Visibility through our three small windows was zero. Our submersible, equipped with numerous and redundant survival and communication systems, had been meticulously engineered to survive anything. We had telephones, antennas, signal balloons; we had enough power, food, and oxygen to last for several days. Just about the only situation no one had imagined was the one we were in.

We had no way of communicating with our mother ship on the surface, and even if they had known about our situation, there was nothing they could do to help us. If the damage to our sub was not too great, we could carefully rise out of the trench; but if our engine was damaged, we would slowly run out of oxygen. And it would be several hours until the debris settled sufficiently for us to find out if we were going to live or die.

In my life I've made numerous small and easily correctable mistakes—and several cataclysmic ones that have jeopardized my business and even my life. But nothing had prepared me for this. The three of us sat there mostly in silence, each of us lost in our own thoughts. Our submersible was the most high-tech undersea craft ever built. For all practical purposes it was considered indestructible. So as we sat below the *Titanic* waiting, it was impossible not to appreciate the irony.

When it was launched in 1912, the *Titanic* was the largest and most technologically advanced ocean liner in history. Because White Star Line believed their own boast that the ship was unsinkable, they had failed to equip it with sufficient lifeboats, so when it struck an iceberg and sank on its maiden voyage, more than fifteen hundred people died within hours in the freezing sea. For decades the wreckage had been lost on the ocean floor, but in 1985, with the assistance of the U.S. Navy, oceanographer Robert Ballard discovered it two and a half miles below the surface. For several years Ballard and his team were the only people to explore this remarkable site. So when my friend and business partner Mike McDowell, through his company Deep Oceans Expeditions, offered me the opportunity to dive to it, I immediately accepted. Ours was to be the first completely private expedition to go there, and truthfully we weren't sure what we would find.

I have an almost desperate need for adventure. Reaching remote, inhospitable locations has attracted me since I was a child when I crept into small caves carrying only matches taken from hotel rooms. Being told I couldn't go, or shouldn't go, or was not allowed to go has always piqued my desire to do so. If I feel stuck in a rut, I try to escape from it through an extraordinary experience.

My heroes are people who took epic journeys into the unknown, often at substantial personal risk. I am simply following the path that they carved into history. I fancy myself an admirer of men like polar explorer Ernest Shackleton, though my expeditions do not rise to that level of danger. In fact, it's quite the opposite; before I begin a journey I do a tremendous amount of research about where I'm going and the equipment I'll be using, because danger isn't especially appealing to me. And I didn't actually believe the *Titanic* trip was dangerous.

But the dive had seemed cursed from the start, when legal reasons almost prevented it from happening. Ballard had returned to the site several times since his initial discovery, and was actively trying to prevent anyone else from going there. When Deep Oceans Expeditions also found the wreckage, using publicly available data, Ballard obtained a preliminary injunction from an American court that, basically, prohibited other people from trespassing on the site.

During the court hearing Ballard had displayed a piece of plastic he had found on the wreckage that had fallen off the motor of one of the Deep Ocean Expeditions submersibles, claiming it was evidence that visitors would be disrespectful to the *Titanic* grave site. While Deep Ocean Expeditions did fail to notice this one piece of plastic, we were of the view that Ballard's expeditions had done considerably more damage. We ultimately prevailed in court for a different reason altogether—it was pointed out that the *Titanic* lies in international waters and an American court did not have the jurisdiction to dictate what anyone could or could not do at the site.

So we were finally legally cleared, though much preparation still lay ahead. This is not the kind of expedition for which you just pack a bag and go. I wanted to understand every facet of our incredible machine. By the time we made our voyage, I was such an expert that if my sub commander died, I could literally have guided us back home. I knew this wouldn't be a luxurious experience. The submarine's hull is about six feet in diameter and it seats three people, though not especially comfortably, and is built for functionality: its inch-thick nickel steel can withstand the enormous pressure five thousand meters—more than three miles—below the surface. Subs are regularly pressure tested to well beyond the depth we would be traveling. And if it did implode, well, at least it would be a quick death!

But we knew that engineers had considered the long list of hazards that might be encountered on a deep dive. After being crushed, the next greatest fear most people have is that the sub will get stuck on the ocean floor, and eventually the crew will suffocate. To manage this risk, Deep Ocean Expeditions' *Mir* had three independent power systems and carried several days' worth of oxygen. But if for some reason its motors failed, then the first thing you wanted to do was make it buoyant so it would float to the surface. The easiest way to accomplish that was to pump water out of the ballast tanks. If that failed, two massive external battery trays could be dropped. If that still wasn't sufficient, pressing a button opened an electromagnet that dumped nickel shot weights on the seafloor, further increasing buoyancy. That system was designed to work even if the batteries died. So if something went terribly wrong inside this vehicle and everyone lost consciousness, the batteries would run out, the weights would be dropped, and the sub would pop up to the surface.

Perhaps the greatest danger we faced was becoming entangled in nets or ropes or the wreckage itself. To try to avoid this, every external object on this submersible was designed to be easily ejected. Its two large light booms, two grasping arms, two sample trays, and three external thrusters all can be jettisoned if they get caught on something. Another safety mechanism is a hatch on top that releases a balloon attached to a very, very long rope that rises to the surface, alerting the surface crew that the submersible is in trouble and pinpointing its location. The submarine also carries two line-of-sight acoustic radios, providing primary and backup communications capabilities.

It is an extraordinarily safe machine, and as we descended into the North Atlantic I felt complete confidence in it. The dive itself took almost three hours, and there was little conversation during that time. I was running on adrenaline. Occasionally I would ask a question and one of the crew members would respond. The trickiest maneuver took place at the very beginning, when we were hoisted off the deck of our ship by a crane and rocked back and forth until we dropped into the water. Unlike boats made to float upright and drive through the waves, the *Mir* is a capsule, considerably heavier on its bottom, so when it hits the water, it rolls. Anything not strapped down, including the crew, gets thrown around. If you're prone to seasickness, it's terrible. But within seconds it starts to descend, then stabilizes, and the ride becomes very smooth.

For several minutes after launch, light from the surface still reached through the depths, so we could look through our windows and watch the sea life and our air bubbles rising. After that, the sea grows darker and darker until it becomes black. Only objects in our lights, which extended twenty meters, were visible. Eventually we turned them off to conserve power, and we had descended into the darkest night I had ever seen.

When our instruments told us we were nearing the bottom, we turned our lights back on. Each of us stayed glued to a window, waiting for our first glimpse of the bottom. Suddenly, we were there, raising a cloud of silt, within range of the wreck. In my imagination I saw the remains of the *Titanic* emerging out of it, but even as the sediment finally dissipated I didn't see the ship. We rose a few feet and started moving slowly. The first debris we saw were lumps of coal that more than a century ago had been shoveled onboard with the expectation of powering the great ship to its triumphant arrival in New York. Instead this fuel had ended up on the ocean floor. Then we saw end caps from a deck bench, the skeletons of deck chairs, and some porcelain fixtures that probably came from a toilet. As we moved closer to the wreck, the debris became thicker and the stuff of daily life had been strewn wildly about: broken plates, timber, flooring. Rather than the hallowed, respected, almost pristine site Ballard wanted us to believe we would find, the ocean floor around the wreck was a mess. And then we saw the ship itself.

I had spent many hours looking at film and photographs in anticipation of this moment, but still wasn't prepared for it. Almost four hours after launching, we finally began exploring the wreckage. We glided in respectful silence around the entire hull. We put down on the deck and ate lunch. We descended slowly through the deck into the main staircase and were able to look down one long corridor, watching in awe as the crystal chandeliers, hanging mostly by their power cords, swayed slowly with the currents. Bacteria had eaten away some of the metal hull, which enabled us to look into the crew quarters and even some bathrooms.

Imagine if your own home was flooded from floor to ceiling. It would be much like you had known it; there would be identifiable reference points, but in fact it would be a completely different, almost ethereal environment. Anything that floats would move, but everything attached would remain in place. That was the *Titanic.* For many people it would have been very easy to look at it and hear the orchestra playing somewhere in the distance. But not for me; I had never been caught up in the romance of the great ship on its voyage. Instead, my interest was in the artifact as we found it. I was far more intrigued by seeing remains that had sat on the ocean bottom for almost a century than imagining a party in progress.

We explored the wreckage by sections. Finally we went to see the screws at the stern, the giant propellers that powered the *Titanic* through the ocean. The engineering was state of the art at the time, but it has become clear that among the few things that might have saved the ship that night were bigger screws and/or a larger rudder that might have enabled them to turn the ship to safety. Both of them were undersized for a ship of this stature.

When the *Titanic* sank the stern was the last part to go underwater. When it crashed into the ocean floor and slid, it cut a fairly long channel, a depression with raised "dunes" on either side of it. This channel allowed us to get under the stern by going over a little hump, or berm, and into the depression. We stayed there for a while, looking at the three main screws; it was incredible that they alone had powered this great ship. Finally we prepared to return to the surface.

The *Mir* has an extremely limited field of vision, with just three tiny windows facing forward and downward. There was no way of seeing above or behind us. We lifted off the ground slowly and reversed the main thruster—and suddenly hit the dirt berm directly behind us. In spite of the fact that we were moving slowly, we hit it with enough force to break off our titanium rear bumper. To avoid hitting the berm again, we raised the sub a bit higher—and crashed into the *Titanic*'s stern.

In my memory I can still hear the deep, resonating thud as we hit the hull. The world suddenly turned an ugly gray and black. Debris from the stern rained down on us as we sank back down to the seafloor. Our visibility was reduced to zero. As we settled on the ocean floor, shocked and incredibly surprised, the realization hit us—we were stuck beneath the *Titanic*. We had no way of knowing how much damage had been done to our sub, or how much debris now lay on top of us. It was possible the entire weight of the *Titanic* was now resting on our vehicle.

In other circumstances I might have laughed at the irony: Maybe like so many of those people who built and perished on the *Titanic,* I had been lulled into a sense of security by my confidence in human technology.

My nature in a situation like this is not to panic. Panicking would have been the opposite of useful. Of course, we were concerned; very, very concerned.

In my mind I went through each of our safety systems one by one and realized that each one had been defeated. Given our obstructed location, our two "line of site" acoustic radios were of no use. With the stern directly above us, we couldn't generate the buoyancy to lift off. Even if we released the emergency balloon, it would just bump up against the underside of the ship! There was nothing we could do but wait. If we failed to surface, no one would know what had happened to us for a very long time—if ever. After all of the preparations, all of the precautions, we could be the latest victims of the *Titanic*.

We knew our only option was to wait until the debris settled so we had some field of vision, try to determine if there was serious damage to the sub, then attempt to surface.

We waited. There was little conversation in our vehicle.

We waited. The sediment settled slowly. I learned some important things about myself in those long minutes. I didn't panic, either outwardly or inside—and neither did the two men I was with. If we really were stuck, there was nothing we could do about it. Crying wouldn't help, complaining wouldn't help. In those moments I probably came as close to the core of exploration and adventure than at any other time in my life. The environment was completely in charge.

We waited. Our visibility cleared slowly, but we were still looking into darkness without any way of determining the extent of the damage. Finally it was clear enough for us to find out if we would live or die.

We turned on our motor. It started instantly. Slowly, achingly slowly, we began moving, freely. We were not being held down by the wreck of the *Titanic*. We backed out. Nobody said a word, there was no nervous laughter, we just moved backward until we were clear of the stern, then started rising to the surface.

At that moment it was impossible not to wonder: How did I get into such an improbable place and situation? The truth was that every single thing I had done in my life had inevitably led me there.

2 CREATE

I Was a Teenage Dungeon Master

My first forays into adventure began more modestly, in my parents' garage. In 1977, the summer after my sophomore year in high school, my parents sent me to the University of Oklahoma for seven weeks to study geometry, statistics, and computer programming. While there I learned to play Dungeons & Dragons. D&D is a role-playing game, an interactive story that is negotiated between the narrator, the dungeon master, and the participants, or players. The dungeon master, or gamemaster, creates the elaborate fantasy and the people playing with him or her take the roles of specific characters in that story, interacting with the other players to solve problems and move the adventure forward.

I had just finished reading *The Lord of the Rings* and this game was the perfect vehicle for creating my own interactive stories. When I got back to Houston, my friends and I began our own Dungeons & Dragons group in my house. Our crew grew so rapidly that at times we had four or five groups of as many as a dozen people playing separate scenarios throughout the house. My mother had already turned our two-car garage into her art studio, but we encouraged her to build a new three-car garage with her studio on top—then we took over the original one. Almost every Friday and Saturday night for several years we had numerous scenarios being run simultaneously by excellent gamemasters. People would play with one gamemaster for several weeks then move to another game. Even parents and one of our teachers would come and play regularly.

While this was before the emergence of computer games, already many people believed Dungeons & Dragons was a dangerous game that was literally turning America's children into devil worshippers. Rumors spread that college students playing the game had disappeared and might have been sacrificed. While it was perfectly fine to read or write a book that included wizards, demons, and devils, it was not acceptable to role-play it. People thought that by playing this game we were literally inviting the devil to inhabit our souls and so would be lost to the Kingdom of Heaven forever. I was amazed to discover how dangerous sitting around a table and talking might be.

But I played anyway, and I learned that each game was only as good as the ability of the gamemaster to craft a story and manage the negotiations into a compelling narrative. Over time, we all took turns being gamemaster and it was quickly obvious that while some people were very good at it, most were not. That's when I began to really understand the power and the beauty of storytelling. Some gamemasters seemed to be satisfied with, or incapable of moving beyond, the simple narrative: Here's a room with level-one goblins. Here's a room with level-two goblins. Here's a room with level-three spiders. There was no arc to the story being presented, little interaction, and few opportunities for players to invent creative solutions. It was just a matter of fighting monsters and grinding onward. Their games became about dice rolls or win/lose as opposed to a magical story that was jointly being crafted. These games were not particularly interesting.

Creating stories came naturally to me; in school I'd struggled with English, especially spelling and grammar. So instead of regular English I took creative writing, a course in which I was graded on my ideas rather than my writing mechanics, and I went from Cs to As. Which is how I got through school without learning how to spell or use correct grammar. To this day, even though my career depends on typing, I have never learned to touch-type.

I was seventeen when I wrote "The First Age of Darkness: The Story of Mondain the Wizard." Clearly the influences that would shape my world—and the worlds I would create—were already present. "*Here is the story of the Village of Moon, and the creation of the world Sosaria,*" it began. "*This is the only written history remaining from the first year of Lord British, immortal, twenty-seventh-level wizard, ruler of Sosaria. Before men, elves, dwarves, or the many other creatures of Sosaria were created, the fate of all races was decreed in the heavens and shall remain so until the end of time.*

"Thus begins the story of Moon, first area inhabited by the elf children and men alike, ________, the first leader of the province of the stars, as it is now known, first proved himself by recovering the scepter of the almighty king from Mondain, first known dark lord of all Sosaria . . ."

But I was far from the best storyteller at the D&D gatherings; at first my stories weren't good at all. They invited players to go off and fight monsters, then go fight more fearsome monsters. Watching other gamemasters, I knew their narratives were better thought-through than mine. My friend Bob White, for example, was really into dark, demonic stories, exactly the sort nervous parents feared. He was far better than I was at having his scenarios add up to a metanarrative of darkness and despair; his stories took players through psychological struggles and evil machinations.

I became a better storyteller by observing the strengths and weaknesses of other gamemasters. Some of them leaned much too heavily on numbers crunching, which is just not fun. It may be the fairest way to make sure that the player with the best strategy wins, but it is definitely not the most entertaining. Bob White, though, could both challenge and entertain. While his themes were often darker than I might have chosen myself, he proved to me that it was possible to simultaneously create a strong interactive narrative and a thoroughly enjoyable ambience among the players.

Playing D&D was the seed that would eventually grow into my elaborate computer games. But still, for a while my own D&D stories were rather mechanical. I would set up a moody scene: "You crack open the door to a hallway and in that hallway there are a row of hangman's nooses. A body is hanging from the noose at the end of the hallway. And as you skulk through this hallway the nooses are 'magnetically' attracted to you, and get as close to you as possible." I would try to make my scenes as spooky as I could; if you touched a hangman's noose, suddenly you would be in it; if you touched a dead body, it would come to life. Gradually my stories became more textured; instead of just figuring out how to survive conflict with evildoers, my players were forced to make simple value-based moral decisions. And rather than merely moving from confrontation to confrontation, these stories began to mimic real life.

The more people enjoyed listening to my stories, the more I enjoyed telling them. Every storyteller is familiar with the pleasure that comes from sitting with your friends around a fire, pouring a few drinks, and weaving a yarn. This was man's first form of entertainment, and when done well is still his best. As I learned, when telling a story well you hear it in your own head—and you see it on the faces of your audience.

Much of my life is about telling stories, the real ones I've lived and those that I make up. More than anything else that I am or have accomplished, I am a storyteller. The greatest reward for a story well told is positive feedback, not always praise, but simply the smile on someone's face. And I got that early on. While Dungeons & Dragons was my first formal storytelling experience, in fact I had been telling my friends stories years earlier when we were playing in our tree forts, and I understood even then that encouraging participation was key. Today when I tell a story, I will still pause and ask listeners directly, "What would you do in that circumstance?" I involve them, forcing my audience to put themselves in my position.

While playing D&D I realized that I needed to be a storyteller; I needed to tell the grand interactive stories that connected all of the disparate aspects of my life. So I learned how to use and program rudimentary computers, because this was the best tool for creating bigger and more exciting stories that I could share with my friends. I have also amassed substantial collections of a variety of unusual items, including, for example, the largest collection of automata, or mechanical toys, in the world; a sizable collection of antique arms and armor inspired by my gaming interests; and numerous obscure scientific items like flexible sandstone and lightning-created "fulgurites." I became a collector for a simple reason: Behind every item in my collections is a story. The item enables me to tell while I show. And the same can be said about my adventures; I've looked down at the earth from the International Space Station and up at the bottom of the *Titanic*. I've hunted meteorites in Antarctica and been trapped on a drought-reduced Amazon. And I love to tell these stories.

My father is a very different man. He too has spent time in space, yet is utterly dispassionate when talking about it. When I was in school, kids would find out my father was an astronaut and would ask me what he said about being in space—and I'd have to go home and ask him, because he never really discussed it. I'd have to push him and even then the most he would say was, "Oh, it's comparable to scuba diving." There wasn't one time that he came home after a flight and said, "I have to tell somebody. I just went on the coolest adventure imaginable and I need to tell someone about it." For obvious reasons we used to call him Spock—and he even came to my wedding dressed as Spock.

Me? You can't shut me up. I have to tell people my stories. I am compelled to tell them over and over. If someone approaches me and says, "I know you're probably bored with telling that story about going to the bathroom in space," I'll interrupt that person and say, "No, no I'm not," and launch into it. I love telling my stories because that's my small way of reliving the experience.

Storytelling is an art, and it can be learned. The foundation of a good story is in the details, the little things that most people don't notice but that enrich the entire presentation. I always pay great attention to the details of my environment; when I see an object that I can't identify or seems out of place, I want to know what it is, what it does or how it works and why it's there, and if I don't know, I stop and figure it out. When I first walked the streets of New York City, for example, I noticed that every few blocks stainless steel tanks were chained to a lamppost. Most of the New Yorkers I asked hadn't even noticed them and the few who did had no idea of their purpose. But I needed to know. In fact, I discovered that they protect underground wires in pipes constantly pressurized with nitrogen. In my games every item must actually serve a real function, and in my storytelling it is the inclusion of those details that makes the difference between reporting an event and spinning a yarn.

I try to bring that same attention to detail to my stories, whether I am standing in front of an audience or creating them for players. For example, this is how I tell the story of what happened to me late one night. Every word is true.

The Night Comet Shoemaker Levy 9 Crashed Into Jupiter

It was almost three o'clock in the morning when the sound of a glass door being smashed open woke me up. I was alone in my large Texas home. A few seconds later I heard someone walking on the broken glass. And I felt a chill—the stranger had come back.

Hours earlier Britannia Manor, as my house is known, had been filled with friends who had come to watch huge fragments of the comet Shoemaker-Levy 9 crashing into Jupiter, the first time in history it was possible to watch an extraterrestrial collision in our solar system.

I'd built this house on the highest point in Austin just for nights like this, and its centerpiece is a large telescope. When my friends arrived early in the evening, I had opened the security gates in front of the house and turned off the exterior lighting so it would not interfere with our observation. It was amazing to see the planet-size dark spots come into view as Jupiter rotated, but the spectacle lasted only a few hours, and by ten o'clock my guests were gone. When the doorbell rang just after midnight, I didn't think too much about it, guessing that someone had left something behind or a friend had shown up very late for the party. I'd forgotten to close the front gate.

I went to a window. Standing at the front door, shifting nervously from side to side with his hands jammed in his pockets, was a stranger. He was wearing a baseball hat, a Rolling Stones T-shirt, jeans, and tennis shoes. The hat was pulled down over his eyes.

I didn't answer the door; instead I stood there watching him to see what he was going to do. I stood at the window for about a half hour. Occasionally he would walk around the side of the house, and I moved from window to window to follow him. He seemed innocuous, but he didn't leave. I couldn't figure out what he was doing; then it occurred to me that he was waiting for me to get home. I began wondering how I could encourage him to leave without letting him know I was home. The front gate could be operated remotely. I closed it, and as I had hoped, he looked around, surprised, watching the gate slowly shut. He must have realized he was on the wrong side of the perimeter, because he hopped over the fence and disappeared into the darkness. I watched for several more minutes to see if he would come back. Weird, I thought. But when he didn't return after thirty more minutes, I went back to bed.

Three hours later someone hurled a large rock through the rear glass doors to my indoor pool. It was the stranger. I realized that he must have been standing in the darkness for hours, just watching my house and waiting. A bay window in the master bedroom looks down on the pool; I rushed to this window and looked down to see him cautiously entering my house. If this was simply a robbery, he'd probably waited and watched until he was certain no one was home. If I banged on the window above him, I thought, he would realize somebody was in the house and probably take off. I started banging on the window—and I was so agitated that my fist went right through the double-pane glass. The window shattered and a thousand pieces of glass rained down on the pool beside him. The intruder stopped and looked up at me. For a few seconds we stood like that, just glaring at each other. Then I said to him in a clear and loud voice, "Get the fuck out of my house!" He stood perfectly still for a few more seconds, as if considering his options—and then he walked into the house.

I called 911. I'd worked with the Austin Police Department on several projects, and I had friends on the force who knew me. I'd never asked for help before. The dispatcher told me officers would be there in fifteen minutes. Fifteen minutes! It would take the intruder only a few minutes to find his way to my bedroom.

My gun safe in the bedroom closet had about a dozen weapons in it. But I had those guns for the same reason I had crossbows, battle axes, bows and arrows, even a working cannon; they are all part of the pantheon of collectible history for me; knowing about them is essential for creating games. Until that moment I had never even considered the possibility that I might actually have to use one of those guns to protect myself. None of them were loaded; in fact, I didn't even have ammunition for most of them. There were some loose bullets in the drawer, but I wasn't certain what ammunition fit into which weapon. I had two Glocks but I ignored them; the last time I'd tried to fire one of them I'd apparently used the wrong ammunition because it clicked and clicked but never fired. I couldn't risk that happening again. I had to make sure I had the right clip in the right weapon.

I picked up an Uzi. I pulled back the slide and snapped the clip into place. I could hear him moving around downstairs talking to someone. I hadn't seen a second person, but at this point it was logical to assume he wasn't alone. I still had the police dispatcher on the phone. "He's talking to someone," I whispered. Then I asked, "He knows you're on your way. I need to know my rights; what do I do in this situation?"

The dispatcher answered matter-of-factly, "Mr. Garriott, if you feel threatened inside your own home by an intruder, you shoot him." Just like that, "You shoot him." It was as if one of my stories was coming to life—*in* my life!

The house is designed with entrances on both floors, and all of those entrances are made of glass. If there were two intruders, it would be very difficult for me to know which direction they were coming from. I'd better move, I decided. I walked out of the bedroom and stood at the top of the main staircase, which meant I had a good view of more of the house.

A few seconds later I heard footsteps crunching on the broken glass directly below me. Then the intruder started walking up the stairs. He moved slowly, and didn't look up at me for the first few steps. Finally he paused, and for the first time saw the Uzi pointing at his head. I warned him, "Stop right there or I'll shoot."

He stopped. We stood there, maybe six feet apart, staring at each other. I held the gun steady, my finger lightly touching the trigger. At that distance I wouldn't miss. I was still on the phone with the dispatcher. I described the intruder to the dispatcher and as I did I realized that he was dressed very differently from the person who had been at my door hours earlier. This man had no cap, no Rolling Stones T-shirt, not even shoes on his feet. That meant there were probably more people in my home! One of them might have gotten behind me. Somehow my already overwhelming sense of dread grew even stronger.

"What's going on, Mr. Garriott?" the dispatcher asked.

"He's still standing in front of me. I'm aiming my gun right at him. This man is not dressed the same as the man who was outside my house!"

"The officers are on their way. They'll be there in less than ten minutes." Ten minutes seemed like an eternity. We stood there, waiting, staring silently at each other for at least five minutes; then he sighed, shrugged his shoulders, turned, and started walking back down the stairs.

When I tell this story, this is the place where I always pause and ask, "What would you do in this situation? Please answer right this second, what would you do? You've got an instant to make a decision. Anything you do at that moment, from shooting him to absolutely nothing, would have been both understandable and forgivable. So which is it?"

In reality I remember thinking, I don't want to kill this person, but it would be dangerous to let him walk away thinking I wouldn't fire my weapon. That would be an invitation to return. So I aimed the gun just a few inches to his side and fired. I blew a hole right through the wall, dangerously close to him. He didn't even flinch. He just continued walking, his back to me.

"What happened?" the dispatcher asked urgently. "Mr. Garriott, what happened?"

"It's okay. I fired a warning shot." The intruder walked out of the room and I lost sight of him, but I could hear him walking around, once again talking to someone. I stood there, frozen in place.

The police finally arrived. When I opened the door, they rushed inside, guns drawn. Two of them went upstairs, two other officers searched downstairs. They found him in a guest bedroom, sitting nearly naked on the edge of a bed. He didn't resist.

The officers handcuffed him and ordered him to lie facedown on the floor. They told me, "Stand here and guard him; if he moves, you shoot him," while they continued searching the rest of the house. I did as I was told and he didn't move. Nothing about the entire incident seemed to affect him. I am still unsure about what I would have done if he'd moved.

He was alone. When the officers returned they sat him up and began questioning him. It immediately became clear that he was very troubled. His name was Daniel Dukes, he explained, and he told them that he had seen a hologram over my house of me beckoning him there to receive the reward he'd earned for completing his quest. With that, he started describing quests as scenes out of my games or the haunted houses I'd created. In fact, he claimed, once he had gotten inside the house, I'd trapped him there and he was actually trying to leave but couldn't figure out how to escape.

The police placed him under arrest. His parents told law enforcement that he'd suffered from a mental disorder for a long time and that they had given up on him. The police detained him as long as legally possible, then released him. When they did they called to warn me to be careful. The police arrested him several more times the following year and let me know every time, but eventually he seemed to have drifted away.

Several years later his obituary appeared in the newspaper. Daniel Dukes had died at SeaWorld. Initially authorities believed he had jumped into the killer whale tank and been bitten by Tilikum, who was known to the public as the first Shamu, and died.

But they found his camera and the undeveloped photographs told a different story. He had been hiding in the bushes there for several days before jumping into the tank, essentially taking photographs of women's backsides. He had died from hypothermia; apparently he had not been injured. The fact that he was found on the back of Tilikum led authorities to speculate that the whale had recognized him as an air breather and might have been trying to save him.

And that's how I learned to tell a story.

3 CREATE

The Origin of Origin

The first video game I played was *Pong.* I was eleven years old when my friends and I wired the game into the back of our vacuum-tube console TV in the family room and began playing. It was magical; you could bounce this square dot off the sidewalls or just hit it back and forth. This was the simplest computer game imaginable. But being able to play a game on the family TV was astonishing. It turned the set from a passive receiver into an interactive terminal. It was a wholly new form of entertainment. I had no concept of whether the technology was simple or complex because it seemed so far beyond my own skills. The idea that I might create my own games had never occurred to me; I had never seen a computer of any kind, and personal computers did not even exist! But I was intrigued.

I was a Cub Scout, Boy Scout, Explorer Scout, and member of Junior Achievement. Although I spent my life around plenty of people with plastic pocket pen holders we were too young to officially be considered nerds. It would be more accurate to describe us as protonerds.

I took things apart, I took everything apart—and *most* of the time I was able to put them back together. I was considered the scourge of the family in anything electrical or mechanical. Anything that came into the house with motors, gears, or switches in it became my victim. At that time Legos and Tinker Toys were not designed to be motorized, but I was always trying to put a small motor in my constructions. Of course that motor had to come from somewhere. One of my oldest and deeply prized possessions is a motorized submarine I built out of Legos. This submarine has a hollowed-out center that holds two double-A batteries. I took an electric motor out of a slot car, I found a propeller somewhere, and I glued the whole thing together so it could actually motor around in the bathtub with me. But when I started building bigger machines, the little 1½-volt slot car motors were insufficient. I needed more power.

"I need more power" is not a statement that most parents like to hear. I opened up our brand-new record player and learned that the band Alvin and the Chipmunks recorded their songs at 16¼ rpms but playing them back at 33⅓ gave them that high-pitched sound; equally important, I learned how the gearing structure worked, how the mechanism moved the gears on the motor and the central shaft that turned the turntable. Once I understood that, I took the motor out of the record player and put it into my own creation.

Sadly for my family, I was not able to figure out how to put it back in the record player.

That was not unusual; I sometimes left behind the parts that resulted from my curiosity. It actually was possible to go through our house and point out where certain experiments had been attempted or where I'd tried to build something. There were places in the house where the carpet not only was stained, but it actually had been eaten through by chemicals. Another decorating tragedy befell our opaque projector, which could project images on the wall—I used my own animation or photos and illustrations I'd cut out of magazines to create my fantasies. But that projector had a lightbulb in it as well as an opening on the bottom to let some heat out, making it sort of like an Easy-Bake oven for carpeting. It went past Fahrenheit 451 pretty quickly and scorched the carpet. At least I never actually set the house on fire.

Well, it's more accurate to say I never burned it down.

I was always exploring, building, experimenting. I wanted to know how things worked, what things did, how the world itself worked. It was a learning time and I did all the experiments children should not do. For example, my friends and I tried to make our own gunpowder, which was mostly smoke. And I was absolutely fascinated by the properties of lighter fluid. We did all those experiments that can be described loosely as: Don't try this at home. We discovered that if we put lighter fluid on concrete and lit it, we could touch it without getting burned because we were in and out of it so quickly. Then we discovered we could put a few drops on the tip of a finger and light it and it wouldn't hurt, eventually building up to soaking our entire hand and lighting it. The lighter fluid evaporates at such a rapid rate that it keeps the skin cool; only after you burn through the final layer of fluid does the heat touch the skin and that lasts only a very brief moment before you blow it out. As long as you held your hand away from your body and swung it down to blow out the flame, there was no harm done.

It All Begins to Compute

My childhood was filled with the pursuit of achievement pins and badges, most of which I proudly still display in my office. At one point the members of my Dungeons & Dragons group formed a Boy Scout Explorer post, and this was where the lines between my various interests began to blur. I suspect it was my mother, our den mother, who suggested we form a post to learn about computers. The only place we knew with computers was NASA, and with my father being an astronaut we had strong connections there, so we solicited the support of one of the space agency's prime contractors, Lockheed-Martin. We would meet at its facility, which was only a few blocks from my house, just outside NASA's gates. They allowed us to use their computer room, which is a pretty good indication of how lightly regarded those rudimentary computers were: one of our nation's leading defense industry contractors, a key NASA partner, permitted a group of teenagers to play with its computers.

Most of their computers ran Fortran, an early programming language that relied on punch cards. Each card was a line of code, and a program was created by using a keyboard to punch holes in cards. A pile of cards taken together was a program, and if you dropped the pile, your program was scrambled. But the room also held one exceptionally magical CRT, the first computer I had seen with a video-display terminal. It ran two programs. One was *Tesseract,* which allowed you to project a cube in three dimensions and play with it. To me, this piece of visualization magic suggested what computers would one day be able to do. The second program was a game called *Adventure.*

Adventure had no graphics, only text, and it forced you to make decisions: "To the north you see the lights of a town. To the south you see a sign that says: To the dungeon. Which way do you go?" Naturally, north. And when you do: "You're in a small town. On the east side of the road is a tavern. On the west side of the road is a mysterious house. What do you do?" That's all the game did. Each location had a text blurb: "You're in front of the house. You see a front door. You also see a mailbox and a doormat. What do you do?" If the player tried to open the door, they might discover it was locked; if they opened the mailbox they might find a letter. If the player decided to open the letter, he or she might be told, "You find a key." The player then typed in directions like "Use the key to open the locked door." By exploring the scene and collecting items, the player was drawn further into the adventure.

There was absolutely nothing to look at, but I was mesmerized. I could visualize everything in my mind. After *Adventure,* I found other text games by luck; there was no place to buy them. Mostly they were spread virally on the so-called sneaker net, which meant that somebody copied a game from one computer on a disc and loaded it onto another computer. Or copies of the games were passed from hand to hand. These games traveled by word of mouth, shared informally among the very few people who actually had their own computer. Even as computer stores began to hang games in plastic bags on a Peg-Board, and distributors began selling them, piracy far outpaced legitimate sales, though nobody thought of it as stealing. These games weren't widely available, so for many players that was the only way to get hold of them. Even the creators didn't consider it a business.

When I began making my own games, there was no economic incentive. It never occurred to me that this could be a career; I was just a curious kid. But two things came together at about the same time: I learned how to play Dungeons & Dragons, and I began writing small pieces of software—and suddenly it occurred to me that I could combine the two. At first, writing games was just another way to tell my own D&D stories. I wanted to make games that would be fun for me to play.

I discovered another computer at my high school, or rather terminals to a computer. They were two teletypes, connected via an acoustic modem to a computer somewhere else. Nobody used the machines, or even knew why they were there. They seemed to have just appeared one day. But they beckoned to me. I convinced the teaching staff to let me have access to them for one period each day. I would work on a project (my games) and they would monitor my progress and give me whatever grade they saw fit. They agreed that we would use this "class," in which I taught myself computer BASIC, to fulfill my foreign language requirement.

I had to use both teletypes. I would type code on one of them, which would punch holes in a paper tape. The second teletype had the acoustic modem, meaning it was connected by phone to a computer at a university. I would dial a certain number and put the phone in a cradle, and when it was connected I would run my tape through it very quickly, which enabled the university computer to run my program. I had no screen, but if I had typed in the right code, it would print out something on paper. It was slow and it was noisy, but even then, to me it was magical. There was something so thrilling about causing a single dot to appear in the place you wanted it to be.

The first words I wrote were the standard first line in every programming manual: HELLO WORLD.

I learned through trial and error, mostly error. At that time there was no computer industry, and there were no standards. In terms of computer language, it was like being at the UN without a translator. A few magazines published programs people had written for the specific computers they were using. It would be a twenty-line program that would add numbers or balance a checkbook or produce a sine wave. Some of them were written in an exotic text called assembly language, and sometimes they were in a version of BASIC. But there wasn't even a universal BASIC. Bill Gates, for example, wrote his own BASIC that was put on an Apple. Another version of the language was unique to teletypes, and there were many other versions available as well. But all the languages used the same general commands, so I was able to dig through the pile and adapt elements of each to my system. That included traditional prompts like print and goto. I would type in these commands, but often they wouldn't work until I figured out how to tailor them to my teletype. When it failed I'd have to sit back and wonder, Why did this crap out in the middle? It made it past line five because it printed out everything before that, but then it locked up. I'd poke around and make some changes to the formatting and eureka! I'd discover that this computer wants this formatting, which I had never read anywhere. It was all trial and error.

I did my junior and senior science projects on the teletype. My father was an upper-atmosphere physicist who studied wave propagation and communication with satellites in the ionosphere. Radio waves normally can only communicate with something within their line of sight, which is why we have relay towers across the country. But under certain conditions it is also possible to bounce radio waves off the ionosphere to communicate with people over the horizon. Before computers existed it was difficult to determine precisely what frequency to use when transmitting and at what angle the antenna should be focused. That early generation of NASA scientists literally used slide rules to make those calculations, but it was largely guesswork. I wondered how—not if, but how—I could do that on a computer. And with a great deal of guidance and advice from my father on the physics involved, I wrote a program that solved that problem. It took the prevailing ionospheric conditions into consideration and showed graphically where the radio waves would go.

My junior year project was called "Radio Wave Propagation with Computer Analysis." That same approach could solve a variety of problems, including underwater sonography and a technique in which people looked for oil and other deposits by setting off explosions and tracking how sound waves bounced off rock formations. That also became my senior year project and it won a lot of awards, including fourth place in an international competition.

So I knew I could program a computer to do complex equations; the more important question to me was whether I could create entirely new worlds! My real objective was crafting a role-playing game. I called my first game D&D 1, an obvious nod to Dungeons & Dragons.

I wrote twenty-eight programs on that teletype. I know there were exactly twenty-eight because I named them D&D 1,

D&D 2, and so on until I reached my twenty-eighth attempt. Many of them were never finished; I'd get halfway through a game and would have learned so much about the process that I would decide to start over, this time utilizing my newly obtained skills to create an even better structure. I never let anyone else play these games. They couldn't; the only way to play the game was to be in the classroom with me, and there was never anybody else in that classroom. So I was not only the entire creative and production team, I was also the entire audience. Creating the games was a laborious process. I had to write the entire program in a notebook before I could type in the code, and there was no way of testing the program until it was typed in. Sometimes it worked, sometimes it didn't. Sometimes the game turned out to be better than my last effort, sometimes it didn't. When a code didn't work, I would debug it, figure out where I'd made a mistake, and try again.

In 1979, by the time I was making D&D 28, I was working at ComputerLand, the first computer store I had ever seen. We didn't have much of an inventory; we sold the Commodore 64, Apple IIs, and the seemingly handmade Sol-20, which was encased in oiled walnut that had been salvaged from a manufacturer of gun stocks and could be bought fully assembled or as a kit. These machines sold for as much as $3,000. They had considerably less computing power than a drugstore cell phone has today, and the limited amount of available software was terrible.

I was a good salesman, but it was a hard sell. These computers existed before people understood why they might need one. They didn't do anything new or remarkable, they just did things that could already be done by hand or on other machines, but maybe a little more quickly. It's a very good word processor, we would tell potential customers, you can move entire paragraphs around—which made it essentially a very expensive typewriter. It could help you manage your recipe card file. While it couldn't handle a business spreadsheet, it could be used to balance a simple checkbook. Manufacturers were boasting that eventually you would be able to use it to open and close your garage door. That we were able to sell any of them at $3,000 a pop was amazing.

The real benefit of working at ComputerLand was that I had access to an Apple II, and I was able to write my games on it. This meant that for the first time other people could play my games while I worked on them at the store, and clearly they enjoyed them. After playing one, the owner of the store, John Mayer, told me, "Richard, this game you've created that we're all playing in the back of the store is obviously a more compelling reason to have one of these machines than anything that's out there. We really need to be selling this on the store wall."

Selling? Wow, what an interesting idea. Most gamers knew that eventually games would be produced for sale; it just hadn't occurred to me that it would happen so soon. Or that it might include my games. Some of the same games that had been handed casually from player to player were beginning to show up in a plastic bag on a hook with a price tag attached. It took some time for people to get used to paying for something they had previously gotten for free. When I wasn't busy making my own games, I was playing games I'd mostly gotten for free, so it seemed kind of strange to me too!

The first game I made for Apple II was named *Akalabeth,* "A game of cunning, fantasy, and danger." I had unconsciously borrowed the name from a Tolkien story, although I accidentally misspelled it: it should have been *Akallabeth.* My sister-in-law had given me that book, which was by far the longest, most complex, and most difficult book I had ever read. I was enraptured. I then read *The Hobbit* and *The Silmarillion.* J. R. R. Tolkien had changed my life. In fact, the name Akalabeth is a derivative of a name used by Tolkien in *The Silmarillion.* I read other fantasy novels, but none of them affected me like Tolkien. They seemed overstructured, as though every time a character was in a difficult situation something magical happened, as opposed to establishing a whimsical world with rules that had to be followed and which required substantial ingenuity from the characters. I would pick up a book and read one chapter and realize it was a poor imitation of Tolkien then put it down in disgust.

Akalabeth wasn't much of a story: go kill a monster. And when the player came back to the castle after killing the monster, he would be told, go kill the next monster. There were ten different monsters, and that was it. Players also were required to go to a town to buy food, or equipment like swords and shields, or to find the only magic item in the game, an amulet that could turn the player into a variety of different beings. The game was not at all sophisticated and there were tons of bugs. But I hadn't created it to sell it, only to prove to myself I could make a decent game. I gave copies to my D&D friends who thought it was cool, but probably not as interesting as tabletop gaming. Nobody ever told me that this was the future. At best they thought it was an amusing way to pass the time until we could get together to play D&D for real.

Akalabeth was basically the computer version of a D&D story; its graphics were very simple. I didn't seek to create the grandest, most dangerous monster ever seen, but rather something recognizable. Visual representation was a challenge; although my mother had tried to teach me art, I had little talent. I needed to fashion something that looked like a monster from the limited graphic tools then available. I had kept myself to about ten line segments to draw it because using any more would slow down the game so much as to make it unplayable. First I made a snake because that was easiest to draw. Then I made a giant pear-shaped rat. I made a skeleton because it could be done with straight lines and was easily recognizable. As with everything else, I had no great plan: the monsters were created based only on how simply I could create a silhouette.

Despite the technical simplicity, there were several elements in that game that were unique, including first-person role-play in several segments; the necessity of going into the town to get food to survive; and the introduction of my game character, Lord British, the ruler who sent players on their quests. The game was so personal, and our expectations so small, that I included my home address and phone number and asked players to call me when they finished the game.

It cost my entire life savings—the $200 I had earned working at ComputerLand—to produce copies of it. That included buying the Ziploc bags. My mother drew the cover art. The copy shop next door printed the instructions and my family sat at the kitchen table and stapled the pages together, then put everything in the Ziploc bags. This was a state-of-the-art operation then. We hung them up in the store and in the first week sold about twelve copies at $20 each. I would estimate that at the time, there were probably fewer than a couple of dozen people anywhere in the world creating computer games, and not one of us could have imagined we were creating an industry that in less than three decades would become the largest and most successful entertainment industry in history, that a game would gross more in a few weeks than the most successful movie in history had earned in decades.

Money Games

Within weeks of *Akalabeth* being "published" I got a call from California Pacific Computer Company. I knew who they were. They'd published games by Bill Budge, a gaming pioneer, whom I greatly admired. I had seen Budge's incredible graphic demos and wondered how it was possible to create those images on an Apple II. Even five years later, when my games were tremendously successfully, I had not come close to doing some of the things he put together for the demos of his games.

I never found out how California Pacific had learned about *Akalabeth,* but they told me they wanted to distribute the game nationally: "We need you to come out to California right away. We've already made reservations for you." For me? I was flabbergasted. Yeah, sure, of course, I said. I was eighteen years old, and only a few weeks after making a game for fun and hanging it on a Peg-Board thinking we might sell a few copies, I was being summoned to California. My parents, who didn't understand at all why this was becoming such a big deal, thought it would be a good learning experience. This was the first time I'd traveled by myself for "business." When I arrived, I was met me at the airport in Davis, California, by a guy in a new DeLorean. Rather than driving back to the California Pacific office, we went directly to an apartment near the airport.

It was a nice apartment in a nice building on a nice street. It belonged to a friend of the California Pacific owner, I was told. In the living room there was a chest of drawers. The so-called friend opened the top drawer and I was stunned to see it was filled, from edge to edge and front to back, with plastic-wrapped bricks of cocaine. Growing up in Houston, Texas, in a community of scientists, I had never been exposed to drug use of any kind; our family generally didn't even have beer or wine around.

Welcome to the video game industry.

Within weeks *Akalabeth* was being distributed nationally. "Welcome, foolish mortal, into the world of *Akalabeth*!" the introduction read. "Herein thou shalt find grand adventure! Created by Lord British."

Lord British, who has since become an iconic figure in gaming history, was another product of that summer program in Oklahoma. When I arrived I was greeted by people who had already been there a few days. They said, "Hi," I said, "Hello." One of them responded, apparently mistaking my NASA neighborhood accent with its Texas twang for what they believed must be a British accent, "'Hello'? Nobody from around here says 'hello.' You must be from England. So from now on we'll call you British." So while Lord British was born in Oklahoma, in fact the nickname was apt: I actually had been born in England while my parents were spending a year at Cambridge. I was born about two months before they returned to America, hardly long enough to pick up a British accent. But the nickname British stuck, and it was so strange but oddly fitting that I liked it. I began using it to represent my character in my games. I always put myself into my games because that was the tradition in my D&D group, and eventually "British" evolved into Lord British.

I credited *Akalabeth* to both Richard Garriott and Lord British, which felt accurate because when I wrote my stories, I would often put myself in my character's place; as Lord British what would I like my castle to look like? What would be my next adventure? Richard Garriott programmed the game, but Lord British contributed mightily in the creation of the story. The people at California Pacific decided, "There's nothing wrong with Richard Garriott, but we don't think it's helpful for marketing. But we love the name Lord British." So the only name on the game when it was published was Lord British. And rather than players calling me when they had defeated the monsters, they were directed to "report [their] feat to Lord British" at California Pacific Computer Company, and in return they would receive a congratulatory certificate signed personally by the good lord.

When the game was published in the fall of 1980, it very quickly became successful and people began wondering about the true identity of Lord British. Among the pioneers of the gaming industry were Al Tommervik and Margo Comstock, who founded *Softalk,* a magazine for Apple II users. They were wonderful people, passionate and enthusiastic about all things within the growing game industry—and unabashed nerds. They were long-haired Hobbits, whose home in North Hollywood became the meeting place for all of us. Their pet parrots were given complete run of the house, so when you were there it felt as if you were inside a birdcage: there was a wonderful sense of being part of nature, although admittedly you also had to be very careful where you sat down.

With California Pacific they ran a contest in *Softalk* offering a prize to the reader who could figure out the true identity of Lord British. They gave no clues other than the name of the game, *Akalabeth,* which meant nothing to anyone who didn't already know me. But people guessed anyway. One example was unusually clever: The *Aka* in the title stood for the well-known phrase "also known as," *la* of course is Los Angeles, and *beth* must be the creator's name. Therefore it was a woman named Elizabeth who lived in Los Angeles. It was a great answer, even if completely wrong.

California Pacific's version of *Akalabeth* was priced at $34, of which I received $5; and they sold thirty thousand copies. I had earned $150,000, more than twice my father's yearly salary as an astronaut. It was a phenomenal amount of money, enough to buy a house. It was so much money that it didn't really sink in; it all seemed like some kind of fantasy. We all thought it was a fluke, though we hoped to extend this lucky streak just a little longer. I decided to trade in the Subaru hatchback I had inherited from my brother Robert for a Mitsubishi Starion, a rear-wheel-drive turbocharged hatchback sports car.

Unlike a singer or an actor who becomes successful after years of preparation and expectation, I had no preparation. It was great that someone wanted to pay me for doing what I was already doing. I would like to say that earning $150,000 as a high school senior didn't change me at all. I would like to say it, but it isn't true.

I was by no means frugal with this newfound money, but other than the car, I didn't know what do with it. None of my friends had that kind of money. I didn't have, for lack of a better term, aspirations of ownership. That sickness would set in later, when I was living in Austin.

I had a very close group of friends with whom I would do most of my medieval recreation on weekends. I cared deeply about these fellow Renaissance nerds, so much so that I patterned the main characters in *Ultima* after them. By that time I was making quite a bit of money and tossing it around pretty freely, probably more openly than was wise. Finally that close group of my friends forced me to sit down with them and told me, "Richard, the three of us have been talking, and you know what, we've all begun to notice that you're becoming something of an asshole."

That was actually the best part; unfortunately they continued, "The money is changing your personality in a way that none of us like. And if you keep going down this path, we're not interested in being friends with you."

At first I was just pissed off. Here I was sharing the bounty of what I'd earned, and now my closest friends were turning my generosity against me. What a bunch of jerks. I was shocked. I was horrified. It was outrageous that they would throw this at my feet when all I was doing was enjoying my wealth and believing I was sharing it with them. How dare they do that? Being rejected by your friends is painful—it hurts.

I heard them, but I was so upset I rejected it completely. It took me a little while to begin reflecting on what they'd said. And the more I thought about it, the more I began to understand what they were trying to tell me. Once the anger dissipated and I started thinking about it, I wondered exactly what I'd done to give them this impression. What things had I done that they had so completely misinterpreted?

Eventually I reached the inescapable conclusion that they were right. I was becoming exactly the kind of person even I disliked. I was headed in the wrong direction; I had been changing and not necessarily for the better. I had a pretty inflated view of my own worth and status and apparently it showed in the way I related to my closest friends.

There was no single thing that had brought them to this point, and most of it was subtle. I was always trying to be the dominant personality in our group, to be the one who shined, the one who got the attention. I was always bringing something to the group that no one else could afford; I would bring expensive liquor to a dinner party, then behave as if that made me the leader. Basically they were telling me they were tired of me constantly showing off, always trying to be the big man. They were telling me to change or take a hike.

It was a very important and pivotal moment in my life. Until that moment I would have described my life as easy. I'd never had a time when I even thought about money. I'd never needed money or had to make decisions based on money. I never had to call my parents in desperation to ask for money. I literally went from not knowing that money would ever be important to being in a situation where I had more money than was necessary or even appropriate. I completely lacked the knowledge or experience necessary to understand the real value or the proper use of money.

In my memory I was able to make the necessary changes pretty quickly to keep my friends, but in reality it probably took a long time. Truthfully, this is something I've struggled with my whole life, but at least I have a much better sense of it now than I did in the past. And I am very aware of the impression that I make on other people and try hard to be sensitive to their needs and their feelings. I can still be very stubborn though. I can still be very opinionated, and I'm pretty happy to voice my opinion at the appropriate time, but I don't have to be the leader, I don't have to be in charge. I may get frustrated when people disagree with me, but I won't stop an opinion different from mine being adopted if that's what the team decides. All I ask is that there be someone in charge who has a clear vision and the ability to describe it and enlist other people, but if there isn't, I'll step up. I've learned how to be a team player. Once a true decision has been made, I won't keep muddying the water with my sometimes controversial counteropinion.

But I had to grow into my success; I had to figure out who I was and how to be comfortable being that person.

There were some difficult lessons along the way. That $150,000 return on a $200 investment was phenomenally high; every game since then has had a much lower return on its investment. While I don't think my parents believed creating computer games would be my career, they encouraged me to chase this windfall for as long as it lasted. We agreed that when it ended, I could get a real job.

I immediately started work on my next game. Initially it was called *Ultimatum*. That was a completely random choice, like most of the names I choose. I liked the feel of it; it sounded badass. The name had no bearing at all on the plot—there was no plot at the beginning—and eventually it was shortened to *Ultima*. The major departure from *Akalabeth* was that I wanted to give players something more engaging than simply killing bigger and bigger monsters. I wanted to put more meat on *Akalabeth*'s bones, and advances in technology made it possible. While *Akalabeth* was essentially a programming exercise that took me about seven weeks of after-school time to write, *Ultima* was my first virtual world, complete with a rich tapestry of characters and activities, and it took about nine months of concerted effort to create. For the gaming industry this was almost revolutionary; most other games took no more than two months to develop.

After *Ultima,* the available memory increased from 48K of RAM to 64K and the game could be written on a disc drive, which provided the space to create lots more monsters—and also enabled players to move around various locations much more easily. That meant I could add a series of activities that were rich and diverse. I added towns and dungeons, and characters other than Lord British who would provide information leading to many different possible adventures. *Ultima I: The First Age of*

Darkness outsold *Akalabeth* by a substantial margin, enabling me once again to put off the pursuit of my "real" career, whatever that was to be.

Risky Business

I was twenty-one years old and right in the center of a brand-new industry. It was like being a pioneer with a newly discovered world to explore, a world where anything and everything was possible. There were no traditions, no rules, no artifacts to collect and honor. We could be as innovative and creative as the expanding technology and our own abilities would allow. It was an industry that attracted all types of wonderful characters, though it also had its darker side.

It still amazes me how pervasive drugs were throughout the industry, especially in California. Oddly, drug use wasn't so prevalent among the creators of the games—we were the nerds sitting in our rooms—as it was among the people selling them. I thought it was insane; no matter how incredibly cool these people thought they were, I was aware that it was still a felony. And the drug use began to take its toll. When California Pacific quit paying his developers, including me and Bill Budge, I heard it was because of the owner's cocaine habit. Ironically, California Pacific offered me the company DeLorean as payment if I would continue developing games for them—at just about the same time John DeLorean himself was arrested for selling drugs.

So I shopped for another publisher. Having created two hit games, just about every company in the industry was interested in signing me—until they heard my demands. Though developers like me were bursting with talent and ideas, publishers were not especially interested in being creative. They all wanted to do the same thing that had been successful—but to do it bigger and better. I'm not sure that they even understood why people loved playing their games. I wanted to enhance the experience for the player. Up until that time every game, without exception, had been stuffed into a Ziploc bag. My desire was to make the world of my games as real as possible, like Tolkien and C. S. Lewis. To do that I wanted to design the player's experience from his or her first encounter with the game, the moment they looked at the box. Except there was no box.

We needed a package that showed this wasn't simply another computer game, it was a portal into another reality. Players needed to believe this world existed somewhere beyond a floppy disc. In addition to putting the game in a box, I wanted to include a detailed cloth map of the world; I wanted instructional manuals that did not give directions like "insert the disc into drive 1 and press the A button," but rather described the magical history of this world. Maybe I even wanted to add some trinkets. I knew from playing D&D that the greater the personal investment you had in a game, the richer the gameplay experience would be.

I had this great dream. Meanwhile, publishers just wanted to put it in a plastic bag and hang it on a hook—all of them except Ken and Roberta Williams of Sierra On-Line Entertainment, which had published the first graphic text adventure games a couple of years earlier. Among their early hits was the *King's Quest* series, which was paid for by IBM and designed for its PCjr. Sierra later published the award-winning *Leisure Suit Larry in the Land of the Lounge Lizards,* an updated version of the text game *SoftPorn Adventure.* The cover of *SoftPorn* pictured the Williamses in a hot tub on the front porch of their Oakhurst, California, cottage. They invited their developers to stay in dormitory-style cabins on their property, and I spent some time there. On Friday afternoons they would break out the peppermint schnapps and just about everyone would get vomitously drunk. All kinds of celebrities would come through. One week Steve Wozniak might be there, and the next week it would be Richard Kiel, who had played the character Jaws in James Bond films.

A high-ranking employee with close ties to the principals of the company was dealing pot on the side, some of it to coworkers. It wasn't a secret, and again, I just couldn't believe people could be that stupid. After publishing *Ultima II: The Revenge of the Enchantress* in 1982, Sierra also stopped paying me. I couldn't believe I had fallen into that trap again.

By then I knew enough about the computer game industry to understand that it wasn't actually an industry; it was an association of companies run by people who had no more experience than I did and who popped up, published a few games, then disappeared—often because they didn't have competent business partners. So my brother Robert and I decided to start our own company. Unlike me, Robert was well-prepared to create a business; conversely, the only game Robert has ever played is *Ultima III: Exodus* and then only because he felt he should understand exactly what he was selling.

Robert is five years older than I am, and growing up I actually had a much closer relationship with our younger sister, Linda. Robert and I weren't especially close friends then. But he was the right person to call when I was in trouble. While I barely finished my freshman year in college, Robert had graduated from Rice with a double major in engineering and economics, then got his master's degree in engineering from Stanford. During those summers he worked on a team at Texas Instruments that was designing the first 64K dynamic RAM chips.

He then got his master's in business from MIT's Sloan School of Management. Oddly enough, while he was there he worked at a venture capital firm where his main job was to learn as much as possible about investing in entertainment software companies. So he had read the business plans of all the companies in the industry. After the failure of my second publisher, he said to me, "These people are not businesspeople, they're gamers. They don't understand that when you get into financial trouble, the people they can't fail to pay are those who create the value of their product. If they shortchange a distributor, they can always find another one. But if they don't pay the person who makes the product, that person will walk away and they will have no company."

Then Robert said, "Richard, why don't you and I go into business together? At the very least I promise you that when the initial checks come in, I will pay you first." That was more than good enough for me.

We named our company Origin Systems because in addition to games, we thought we might eventually expand into business software, operating systems, video entertainment complexes, or other related products. We did end up creating dual joysticks that worked very well, but a Chinese manufacturer told us it would cost $10,000 just to make a mold, and considerably more to make a finished product. It wasn't clear that there was any market for cool joysticks. I still have a bag of prototypes lying around.

Origin was a good name for a company we hoped would be a fountain of creativity. Robert, in a moment of creative brilliance, created the tagline by which the company became known: "We create worlds." Each of our games was a unique virtual world for players to explore, with a level of detail and consistency that Origin became noted for. This was a time of great innovation; Origin was founded around the same time that Michael Dell founded Dell Computer and Trip Hawkins founded Electronic Arts. Years later, in fact, EA bought Origin.

One of the more exotic prototypes we designed and actually built was a large wooden virtual reality device we called the Nauseator. This was at least a decade before similar—and safer—arcade rides were created. The Nauseator was a four-foot-tall, four-foot-deep, two-foot-wide windowless box made of plywood and four-by-fours and mounted on a large frame constructed from two-by-sixes that allowed it to tumble in all different directions. It was essentially a full-motion space simulator. It filled our entire two-car garage and probably weighed close to a thousand pounds. One person crawled inside through a hatch. Originally there were no seat belts, no way of controlling it, and no monitor or display. Once the hatch was locked, it tumbled freely in all directions. The rider didn't get nauseous inside, but after he came out and tried to stand up it hit him. Our plan was to put a monitor, a motor, and a joystick inside to create a rudimentary three-axis controllable flight simulator. But then we realized how dangerous the device could be. Not only might it fly apart at a good speed, but we had four-by-fours spinning past each other at a very high speed, and if anybody stuck even a finger between those timbers, it would have been sheared right off. So the Nauseator sat in the garage for several months before my father insisted we tear it apart to make room for his car.

We learned from the Nauseator to focus on what we did best: making computer games. We had a great team in place. In addition to Robert and myself, the other founders of Origin were our friends Chuck Bueche, or Chuckles, and Jeff Hillhouse. As a favor, Al Tommervik and Margo Comstock created and ran an ad for us in *Softalk,* "Announcing the existence of Origin. We're the home of Lord British's new game. We're going to be based in Houston, Texas. Come join us on this grand adventure as a planner." It was written in a way that made it difficult to ascertain if we were trying to recruit employees or players.

The company was based in my parents' house. Everything about it was informal. One night, for example, we were sitting in our office, actually my mother's kitchen table, when a car stopped in front of the house and a strange man got out. He appeared oddly disheveled, with long hair and a long beard, and had some kind of furry thing wrapped around his arm. "My name is Dr. Cat," he told us. "I read your ad. I'm here to work. I'm a programmer."

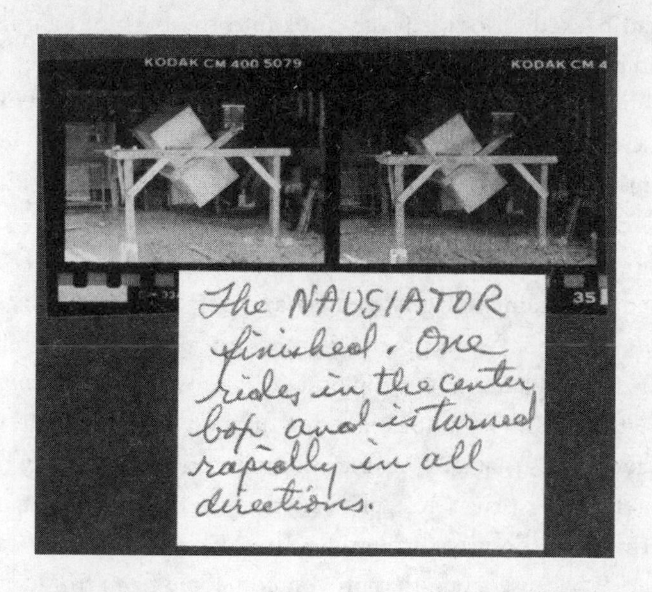

The Nauseator, one of our more outlandish garage constructions.

He had packed all of his possessions into the back of his car and driven down from Wisconsin.

"That's great," we replied. "We need a programmer right now. Come back in the morning and we'll get to work." He was Origin's first outside employee, and he worked for us for fifteen years. But in those pioneering days of the gaming industry, this was not an especially unusual story.

The first game Origin created and shipped was *Ultima III: Exodus* in 1983. While many games are very difficult for the player to complete, most people never discovered that *Ultima III* was actually almost impossible for anybody to complete. The basic formula is that a $40 game should provide about forty hours of entertaining play. Most players make it halfway, or at most, two-thirds of the way through a game. I know that was true of *Ultima III* because a bug that prevented people from finishing the game was discovered several months after it was shipped.

In the game, players were given a clue and told to go and seek out the altar. However, much later in the game they had to ask a character an important question about the altar. And what they couldn't know was that the only way to move past that point in the game was to misspell the word *altar* as a-l-t-e-r, as I did when writing the game. In my defense I have always been a bad speeler, but I am consistent. So when I was testing the game, I speeled it wrong every time. I never noticed and we shipped the game with that error. It was literally months before we discovered it, and there was nothing we could do to correct it. This predated the Internet, so we had no idea who owned the game nor did we have any way of contacting them. But we never received a single complaint from a player that the game was unbeatable.

Even with that terrible error, the game was a big success. I'd managed to come all the way from playing *Pong* to running my own game company. I was very fortunate that I'd grown up into the computer age.

4

EXPLORE

The Wonders of Wondering

There was a dungeon buried deep underground in *Ultima V.* It had a variety of monsters as well as some seemingly helpless children who were locked in separate cages. On the floor was a lever with no indication of what it did or how to use it. But it was positioned in a way to suggest that pulling that lever would release these children, thus saving their lives. We'd put it there believing players would be so curious about it that they wouldn't be able to resist pulling it. In fact, that lever actually opened the gates to the cages, enabling the monsters to eat the children.

This was not the reason for some people accusing me of encouraging child abuse in my games, but since a similar scene in *Ultima IV* did provoke such a reaction, I have included some version of it in every other game I've created because I wanted players to have to stop and speculate about what the result might be before they took a particular action.

There are few stronger motivational tools than curiosity. When I point out to people that the number of times the lateral lines on their palms cross has been shown to have a mathematical relationship to mental health, they can't resist looking at their own hands. They're curious, so they have to look.

Being actively curious has made all the difference in my life.

While everybody is born with some degree of curiosity, my parents always encouraged me to pursue it. They wanted me to be aware of my surroundings and ask questions. My family couldn't simply go on a vacation, for example. If we went to Yellowstone Park, my father wanted us to stick a sterile test tube into some water covered by green slime, seal it, and take it to a university with a good microbe lab to determine what was in it. Years later, when my father and I took a submersible to the hydrothermal vents at the bottom of the ocean, we brought back some samples and had them analyzed; at that time it was believed life on the surface couldn't exist above 100° Celsius. But water does not boil at that temperature under the pressure at great depths and researchers had previously discovered microbes around those vents that lived very happily at temperatures as high as 120° Celsius. The microbes my father and I brought back survived at 130° Celsius, so for about six months we could claim we had found a unique life form, a microbe that would survive at the highest temperatures yet recorded, mostly because we were curious to find out what, if anything, could live at that depth and that temperature.

My mother was responsible for my award-winning kindergarten science project. She took me out into our backyard and said, "Let's follow this bug and see where it goes." That bug was a type of wasp that captures cicadas, paralyzes them with its stinger, and buries them in its burrow, then lays its eggs in the cicada's body. When the larva hatch they eat the inside of the cicada and emerge fully grown. My mother and I actually tracked a wasp around our backyard then dug up one of these tubes with the paralyzed cicada; I built a terrarium and showed the dirt with the cicada trapped in it and turned it into my project.

There is an exercise actors use to make them acutely aware of their environment. Right now, pause and look around you. Indoors or outdoors, it doesn't matter. Take a good look. Then close your eyes and list as many red objects as you noticed.

When you're done, take another look around but this time look for red objects.

The world becomes a different place when you're focused and aware of your surroundings.

Do you wonder about the depth of your own curiosity? Maybe the best way to measure would be to remember how quickly you looked at the palm of your hand when you read that completely fictitious claim at the beginning of this chapter, that the length of your lateral lines relates to mental health.

But sometimes my curiosity got me into some awfully dangerous situations.

I.

I was hanging on for my life, almost a hundred feet off the ground, as my hands slipped slowly out of the crevice. I remember thinking how ridiculous this was; I'd been to the depths of the ocean and was intending to travel in space, I'd trudged across the ice pack in the Antarctic and been marooned on the Amazon River, and here I was about to die just a few miles from my home in Austin.

The cliff was called Enchanted Rock. I had frequently rappelled off it with my friends. Climbing is an exciting way to test your self-confidence but, frankly, as long as the ropes are tied correctly and a few other sensible precautions are taken, it is also quite safe.

This afternoon we were free climbing, climbing without ropes, to the top of a ridge that was easily high enough to kill us if we fell. A diagonal crevice in the rock wall ran all the way from the top to the ground. It was just a few inches across, but wide enough to slip both hands into it and get a firm grip. I watched as my friend Greg Dykes slowly and carefully made his way down the rock, putting hand over hand inside that crevice. He made it look easy, he didn't even use a safety rope, so I figured that if he could do that, I could do it too.

It didn't occur to me that Greg is considerably stronger than I am. Or that I wasn't an especially experienced free climber. So I carefully lowered myself onto the face of the rock, slipped both hands into the crevice, and started moving slowly down the sheer rock face, one hand then the next, a few inches at a time. It was much tougher than I had expected; my hands were supporting all of my body weight. I was about a quarter of the way down when I realized I wasn't going to make it. I just wasn't strong enough. I was stuck; I wasn't strong enough to climb back up to the top or make it all the way to the bottom—but if I let go I was going to die.

I am not a daredevil. I don't push the survival envelope. But I have a need for adventure. I am fulfilled mostly when I am exploring a new world, whether it is in the wilds of Antarctica or a run-down boxing gym. I always want to be doing something new, something different; that's just who I am. And so I visit distant places, jump out of airplanes, rappel down cliffs, test my skills to the edge of my ability. There are people who get their thrills by seeing how close they can come to death, which is not me—the statistics for BASE jumping are far too bad for me to be interested in that activity. I just want to experience all the possibilities that life has to offer. And that is how I happened to find myself clinging to the side of a cliff, trying to figure out how to save my life.

The countdown to my death had begun and I needed to come up with something right then and there. This experience was far more frightening than being trapped under the stern of the *Titanic*. In that situation we were not facing immediate death. There was nothing we could do at that moment but wait. This was an entirely different situation. I literally was hanging on for dear life. I didn't have any time to waste thinking, I was desperate. I discovered that I could jam the forward part of my whole arm into the crack in the rock. While Greg had held on by his fingers, I shoved my forearm up to the elbow into the crevice and used it as a lever to support my body. It was extremely painful, but it was far more desirable than falling to my death.

I hung there for a few seconds and caught my breath. When I felt under control, I slid down another foot, and then another foot, and then I rested again with my arm painfully jammed in the crevice. When I'd recovered my strength a bit, I would let myself down another few inches. It took me just a little less than what seemed like an eternity to get close enough to drop to the ground without substantial risk of injury or death.

This is easily the biggest and potentially deadliest mistake I ever made during an adventure. I learned two lessons: One, I'm not nearly as strong as some of my friends. I am, in fact, a computer nerd. But the second and more important lesson was how quickly a bad decision could manifest itself and put you in a nonrecoverable position. Prior to that experience, while I had observed most safety precautions, I'd probably taken them less seriously than I should have. I had found out that day that I was not invincible and I never forgot it.

II.

My need to understand the way the world works brought me all the way to Antarctica to hunt for meteorites, the same meteorites I almost certainly could have found in my own backyard, or in yours. In fact, meteorites are continually bombarding all areas of the earth all day and all night, every day of the year.

But most of them are very small; by the time they get near the ground they have often either burned up completely or landed as a speck smaller than a grain of sand. When you sit up at night and watch shooting stars, those are often basketball-size meteorites that are entering the earth's atmosphere, and they will leave some small debris for hunters to find. This rain of meteorites is evenly distributed across the earth. On average there is at least one visibly sized meteor lying on the ground per every kilometer on earth.

We generally don't find them because they're hard to pick out among the other rocks and trees and man-made structures. And rain slowly dissolves most meteorites. So even though they may be on the ground for a few years, eventually they get buried or finally rust and crumble. As a result, they are actually quite hard to find in most places.

But there are two environments on earth in which meteorites may still be found in abundance. One is in desert regions like the Sahara, where people search for them with ultralight airplanes or large magnetometers. While they can be quickly buried by the shifting sands, the desert winds may expose them months or years later.

The other region, arguably the best place on earth to find meteorites, is Antarctica. Antarctica is covered with a thick sheet of ice about a mile deep; the only things sticking up through that ice are the tallest mountain ranges. Ironically, it doesn't snow very much in Antarctica, but the ice and snow are distributed fairly evenly over the continent. Meteorites have been falling on the ice for millions of years and have been buried under the snow. Those near the surface might only be ten thousand years old, but the ones buried deep may have struck the earth millions of years ago. As the ice sheet flows toward the coast, it sometimes encounters a mountain range, and when that happens it often burbles up like a wave and then flows around the sides of the mountain. However, during the flowing up, the Antarctic winds will blast away the ice. And all the meteorites that have been trapped below the ice for eons tend to accumulate at that spot. That means that on occasion, when you're walking on the white ice you can actually see the black meteorites lying on the surface.

There is only one private company arranging tours throughout the interior of Antarctica: Adventure Networks International, founded by my business partner Mike McDowell. Going to Antarctica is difficult and sometimes very dangerous; it is expensive as well, so few people make the journey. On the first of my two trips we stayed close to base camp, which was situated at the foot of Mount Vinson, the highest peak in Antarctica, and is essentially a starting point for climbers. But it is too far from the South Pole to be really good for meteorite hunters; as you go north it gets warmer and any rocks that had been on the surface have melted down deep into the ice pack. We didn't find a single meteorite.

I went back two years later on an expedition that went much farther south. By no means was I an experienced Antarctic explorer, or an expedition leader, or even an expert in meteorites; I was as much a client as anyone else, so we hired guides to keep us alive on the ice as well as a geologist to help us strategize and find those elusive meteorites.

Our journey there exposed us to an environment unlike anything I had ever experienced. As I was to see later in my life when I traveled in space, in Antarctica some of the natural laws we take for granted just don't apply. For example, when you pitch your tent, you must stay at least twenty to thirty yards away from other tents. That was because we were camped out on a smooth, flat sheet of solid ice that acted like a wave guide for sound, so even the tiniest noise would travel incredible distances. Against the Arctic quiet, the far-off sound of others snoring is truly disturbing.

Antarctica is literally silent. It is a silence so profound that when the wind stops blowing there is nothing to be heard. There isn't anything comparable. At first you become acutely aware of the little ringing in your ears that we all experience, but after a few days the ringing in your ears stops and suddenly you have superhearing. Then you add that to the wave-guide effect and you can hear incredibly soft sounds that are happening dozens of yards away, but you have no way of determining where they are coming from.

I remember one morning waking up to what sounded like a

fly in my tent. I immediately started looking around for this fly—until I realized that there couldn't be a fly in my tent. There are no insects in Antarctica. Finally I unzipped my tent and looked outside, and the gentlest breeze was blowing; it was not enough to make any noise on my tent, but one of the flaps on a neighboring tent, about thirty yards away, was rubbing against another piece of fabric—the sound I easily mistook for an insect.

Not only couldn't we depend on our sense of hearing to provide the normal cues we needed, in the Antarctic the information usually provided by sight is also unreliable. The absence of common landmarks makes it almost impossible to judge scale or distance. In the interior of Antarctica it is very difficult to tell the difference between small things close up and large things far away. Ice and rocks have a "fractal" structure that adds to this difficulty. One morning we took a small Cessna plane to survey a region for meteorites. We thought we saw a footlocker-size boulder on the ground that could have been a meteorite, so we landed the airplane what appeared to be only a few hundred yards from it. In our normal environment we all know, for example, the size of objects like cars and telephone poles, and we know that as we get farther away from them they become smaller in our field of vision. In Antarctica there are no cars or telephone poles or other man-made objects of any kind. There is nothing to use as a basis for size comparison.

Usually when a hunt begins, people spread out somewhere between five or ten yards apart in a perpendicular line and begin hiking. As we hiked, the plane got farther and farther and farther away, until we were at least half a mile from it—and the boulder wasn't getting any closer. We kept going until the plane was now truly remote, and when we finally reached the boulder, it was the size of a house. It certainly was not a meteorite.

It was on this search that we stumbled onto one of the most extraordinary things I have ever seen. We walked into an area where the wind curling across the top of a mountain had scooped out the ice, creating what looked like a giant hundred-foot-high frozen wave that appeared to be breaking. It stretched for at least a mile in front of us. The top of it was white but the side was more or less clear blue ice, and as we walked under it we could see deep into this wall of ice. I half-expected to see a frozen whale or mammoth or yeti or something caught and frozen there from millions of years ago.

There were two more things we found on the ice, and I will wonder about one of them forever. The first were bamboo rods sticking up out of the ground. We knew what they were: When early expeditions traveled across the ice, they would leave caches of food or other stores behind, or would simply mark their route. If they made it all the way to the South Pole, they wanted to make sure they could retrace their path. So we knew we were on the trail of an historical expedition.

It was the second object that will remain a mystery. As we walked along we found a white glove lying on the ice. Finding anything man-made on the ice at all is rare, but finding such a vital piece of safety equipment is deeply unsettling. It was obvious as we examined this glove that it had been tumbling around on the ice for a long, long time. Modern gloves are made with a fabric shell and fleece or some other polar lining to provide sufficient insulation. Older gloves might have been made of sealskin or some other leather and lined with animal fur. But this was a hand-knit glove, two layers thick, clearly from an old Arctic expedition. It was impossible not to wonder about the fate of the person who had lost it so long ago.

Antarctica is larger than North America, but it is the least populated place on the surface of the earth. As we stood there we knew there were no other humans within thousands of miles of us. It is an environment in which every decision you make can have an impact on your survival. Like space, it is unforgiving.

We were equipped with the most technologically advanced safety gear and still had to be extremely cautious. The other humans who had been there before us had gone through the same—and often much worse—experiences with less gear, so how could anyone have dealt with the loss of such an important piece of safety equipment? I fear that they could not. The glove was the only thing we found. When I think about it I wonder about all the possibilities. But I know this is a mystery that will never be solved.

We made it to the South Pole and along the way found over forty meteorites, ranging from the size of a raisin to the size of a Ping-Pong ball. Initially, meteorites look no different from black rocks, but on closer inspection you can see the crust that develops when they plunge through the earth's atmosphere. Although the elements were always in charge, because we respected the environment and were prepared for it, we were never in any real danger. Seeing the preparations it took and the equipment we carried to ensure our safety made me appreciate even more the courage of the early explorers—and I couldn't help thinking about that knitted glove we'd found on the ice.

III.

Truthfully, I was probably in more danger on the Amazon River. I went there in the early 1990s with no special goal; it was just a place I wanted to explore. The first GPSs had just become available. The technology wasn't very good, but satellite phones didn't exist, so this was a prized safety tool. We were a crew of sixteen, including my sister and future brother-in-law. We came over the mountains of Ecuador, from Quito to the headwaters of the Amazon. Our plan was to go by giant dugout canoe down one tributary and back up another to return to our starting spot. In addition to our supplies we carried a fifty-five-gallon fuel drum to power a small motor if we needed it. The first week of the journey went perfectly according to plan. The second week did not.

The weather there had been unusual. The annual rains hadn't come and the land was very dry, the river getting lower every day. As the water level dropped, debris from the past several decades was exposed. Large trees that had been completely submerged for years suddenly became obstacles. On our trip downstream there had been water from bank to bank, but by the time we started back upstream those trees forced us to slalom between their long branches.

This part of the Amazon consists primarily of long, winding S shapes. A river that has a reasonably deep slope is formed in a more or less straight line. But when it forms on flat terrain, it tends to create these bends. Not only are they winding, but the silt that flows downstream clogs up and dams the river now and then. The shape of those bends changes from year to year and decade to decade. But as a result, the two loops of the S often were so close together we literally could jump on the bank, climb up a short ten-foot sandbank, and look down to the other side of the S curve. By walking ten feet we could cut out half a mile of river; the problem was that our canoe and all our supplies needed to stay on the river. When the motor was working and the water was high, we could make that loop in no time, but as the water level dropped and exposed natural debris, that became more and more difficult.

During those first few days of the return trip, if we got up a good head of steam we could make our way over those tree trunks. But eventually the river dropped so low that we had to get out of our canoe and portage over the logs. And when that became too difficult we had to stop every few hundred yards to chop our way through these obstacles.

The water level continued to drop. Our pace ground to practically nothing. There were days when we made only ten feet of forward progress, and it became obvious that our supplies wouldn't last long enough for us to complete the trip. Our guide didn't know every bend in the river and even he became reliant on my GPS to plan our route. After several days it became clear that we weren't getting out.

It was disconcerting to see what happened as the expedition slowed and our food ran out; we went from passengers sitting around to full-time workers, helping lift and chop to clear a path for the boat. It was painful and somewhat dangerous; many of the trees had sharp spines that were constantly jabbing us, and there were eight-foot-long crocodiles and piranha in the water. Unlike in the movies, piranha, I learned, will not generally attack unprovoked. At first I was reluctant to get into the river when they were in the area, but I followed the lead of our guides, who would bathe in the river every night. It was a little creepy being in the river and seeing them nearby, just watching us; they may have been ignoring us but I certainly wasn't ignoring them.

We progressed very slowly up the river until we came to a place where a tree between four and six feet in diameter blocked our path. We had been able to cut through other trees with our two axes, but this one weighed tons and there was no way we could get past it. We were stuck and isolated. We could no longer move on the river.

As the expedition planner I felt an obligation to try to hold it together, while several others who were with us had grown increasingly unhappy and anxious. Small problems that otherwise would have been overlooked suddenly became issues. There was a sharper tone in people's voices. Relationships were tested as we were forced out of our comfort zones. We even sensed an uneasiness among our guides. The situation seemed surreal. I remained outwardly confident, but even I was beginning to feel some anxiety. Not for myself, but rather for the other people whom I had brought along on this expedition.

Our guides began mapping a route in case we had to walk out of the jungle. That became increasingly likely. We were within a couple of days of being out of food; we were already foraging pretty far into the woods. Perhaps our most immediate problem was that we were running out of fresh water. We watched the skies, waiting for the rains that we knew had to come—and yet it remained dry. Fortunately, the same night we were finally stuck, there was a fairly substantial rainstorm. It didn't raise the water level nearly enough for us to get over that tree, but it gave us plenty of drinking water.

By then we were more than a week behind schedule. Our guide's wife had been waiting for us at the designated arrival point. She was also watching the river and was able to figure out what was happening. She hired two smaller canoes to head downstream. I was doing my best to keep everyone's spirits up, but this was long past being a routine trip down the Amazon. While we believed we weren't actually in danger, we were trapped on a falling river with not nearly enough supplies. Eventually we'd get out of there; the real question was how much suffering we had to go through first.

Soon, I was on to my next adventure.

You Can Do This at Home

You don't have to spend many thousands of dollars to go to Antarctica or the desert to find a meteorite. In fact, you can probably find one within walking distance of wherever you are right this minute. There's likely to be a meteorite as big as a marble within one square mile, something that you can pick up and toss around in your hand. But there are many, many smaller ones. Most houses are pelted with grains of rice-size meteorites all the time. So if you live in a house with a sloped roof and a gutter, go to where the gutter downspout drains onto the ground and look.

To identify a meteorite, run a magnet across the ground around that downspout. It will pick up a bunch of debris, some of it the debris of modern life, such as metal filings from a construction site. But if you get out a magnifying glass or microscope and sort through the magnetic debris, you will find little rocks with the telltale crusting on their edges. Those are meteorites.

So if you happen to be looking for a great science project for young people still in school, this can be it—but please send me pictures or report your discovery to Lord British on Twitter: @RichardGarriott. Each one of my games has included this type of "Report thy feat"—and I always respond!

IV. Austin, Texas

For a person like me who has spent much of his life creating fantasy, reality can hit very hard. As a child I was never a fan of boxing. My family would gather around our classic vacuum-tube TV to watch Muhammad Ali's fights, which were spectacular shows, but other than those rare times I never watched it or, for that matter, even thought about it. It didn't interest me. When I became an adult I liked it even less, considering it a blood sport that often exploited economically disadvantaged young men. I watched the promoters making a bunch of money, but it was rare that a fighter walked away from the sport uninjured and financially secure. So naturally I never expected to find myself working in the corner at a world championship fight. The world of boxing was as foreign to me as trekking through Antarctica. It is a place I never anticipated being, yet once I arrived there, I threw myself completely into it, trying to learn as much as possible and experience as much as I could.

It began several years ago when I was talking to one of my friends about finding some type of exercise program that would be challenging and help me get into good shape. Knowing about my adventurous spirit and competitive edge, he said, "I know about this boxing gym I think you'd actually enjoy." While I was highly skeptical, the concept of actually getting into the ring was strange enough to intrigue me. And also about the last thing I would ever have thought of myself actually doing, which of course is what drew me to Richard Lord's Boxing Gym in Austin. Whatever kind of environment I was expecting, Richard Lord's gym wasn't it. It was essentially a rat hole. It was in the back corner of a corrugated-shed row of buildings, sharing the space with marginal businesses like a tire repair shop. It was a seedy-looking, run-down place. The walls were covered with faded, torn, or water-stained boxing posters from many years earlier. But what struck me immediately was that the people there didn't fit my expectations. While certainly there were some hardened boxers, there were also young women and children, and businessmen who came in wearing obviously expensive suits. It was an array of people from different walks of life, but all of them were walking comfortably among exercise equipment in disrepair, junk piled in the corners, duct-taped mats, and boxing apparatus. The whole scene was surreal. Okay, I thought, this is definitely interesting. I'll give it a chance.

Among the fighters training there were eventual world champions Jesús "El Matador" Chávez, Armando Guerrero, and Anissa "The Assassin" Zamarron, as well as several other boxers who had very successful careers. They trained right alongside—and sometimes with—everybody else in the club.

I began my first workout, assuming I would ease my way into it. But that isn't the way Richard Lord runs his place. Nobody eases into anything. No one cuts anyone else any slack at all. My first day there I was lying on a mat and on either side of me were young boxers. We were doing an endless nightmare of crunches and it didn't take me long to be exhausted. When you're exhausted you tend to stop doing crunches and put your feet down. When I did, the people on either side of me, neither of whom I had ever seen before, became incredibly encouraging, saying, "Come on, come on, you can do this, don't stop. We're not done yet."

I looked at them, thinking, "Who the hell are you? Can't you see that I'm suffering? This is my first time, cut me some slack." But I did a few more, and when I stopped again, other people started encouraging me. Almost immediately I was sucked into this community of people who completely understood and in fact had shared my suffering. Everyone was there to push everyone else to the limits of their capability. Starting on that first day and growing with each visit, I fell deeper and deeper in love with those people, the process, and the gym.

All of it was created by Richard Lord and his wife, Lori. Richard was a fighter and bears those scars. He is a small, sinewy, almost skinny but very powerful man with an enormous depth of experience and unbelievable personal drive. He does each workout with the person he is training, so essentially he works out all day, pretty much every day. He doesn't take breaks; he would eat his lunch while we worked out, literally grabbing bites between punches. He is also among the kindest souls I have ever met; not only does he train and manage fighters, he gives them a place to stay if they need it and makes sure they get to keep the money they earn in the ring. While the gym was such a diverse mix of types, including some people who have come out of prison and others who might eventually be going in, it was all held together by Richard Lord's personality. In some unique way he inspires people to be better.

It did not take long for my entire impression of boxing to change. I used to think boxing was a pretty simple sport; two guys stand toe to toe beating the hell out of each other until one guy can't take it anymore and falls down and the other person wins. Well, that was wrong. Boxing is very much like fencing; it is very difficult to hit someone who knows how not to get hit.

The whole group sparred on Saturday. Richard matched people based on skill and stamina; he would never put anyone into a situation in which they might get hurt. He might take a less skilled person who is fresh and put them in the ring with a slightly better skilled person who is tired. No one was there to hurt anyone else, but it was always a great workout and I always learned something.

It didn't take me more than a couple of sessions to realize that the people in there were not just better than me but a lot better than me. Not just the men of about similar size, I mean everybody; women who were half my reach and weight could rip me apart easily. When we were in the ring sparring, I couldn't lay a finger on them and they could lay me out flat. This is interes . . . Ow! I thought, and my respect for them went up tremendously.

Obviously we wore protective headgear when we sparred, so in fact there was really no danger of being hurt, but having your head rocked is not pleasant, it's like someone throwing a bucket of cold water in your face. It's fair to say that it is extremely attention getting. It takes you out of the moment, and if you're in the ring with a good fighter, they'll take advantage of that and continue to pummel you until you get beyond the point from which you can come back. The natural reaction is to turtle up, protect yourself, and get away from this nightmarish thing that is happening. People around you are yelling to you to punch your way out, but when someone is flailing away at you, it's really difficult to think about regaining the offensive. I had to learn how not to react to getting hit and sometimes getting hurt. Showing pain or distress makes your situation infinitely worse. Ha, you hit me with your best punch and it didn't faze me—in reality, it definitely fazed.

Several female professional fighters worked out at Richard Lord's and he didn't hesitate to put men in the ring to spar with them. Without question, each one was a far greater danger to me than I possibly could be to them. Initially I was reluctant to hit a woman. Not because they were women, but because it appeared that I was taking on someone smaller and frailer than I was. But once we got into the ring, that changed pretty quickly.

Melinda Robinson is a policewoman and professional fighter with a mixed won-lost record. She was about my height and she wasn't quite as fast as some of the other fighters. When I sparred I wore a protective bar across the front of my helmet; while others might call that chicken shit or wimpy, I preferred to call it sanity. It accurately reflected the level of my confidence in the ring. Melinda, though, had an open face mask as she was a very good fighter. And when we were sparring one day, the years of training I'd had finally kicked in; when she dropped her gloves, I tagged her. It was a good hit, square on the nose. She backed off and said to me, "Dude, we're just sparring, did you really need to bean me like that?"

The one time I'd actually hit somebody while sparring, I thought I'd offended her. I'd never hit anyone that well before, so I didn't know how to respond to this. I felt bad about it for a week until I had a conversation with Anissa Zamarron, who from time to time was my trainer. "Oh, dude," she said, "hit her harder next time. You don't feel bad for anyone in the ring. When a fighter wants some slack, they can retire. You're in there to do your job." By the way, Melinda and I are great friends; she's a terrific fighter as well as friend, as is true for everyone else in the gym.

Jesús Chávez was by far the best fighter in the gym. His rise through the ranks was rapid; he won most every fight by a knockout. He began getting tougher fights out of town. If he was fighting nearby, in San Antonio, for example, many people from our club would pile into cars and go there to cheer for him or work in his corner. When we traveled, we would smuggle giant caches of red cloth squares and matador horns that we'd ordered from cheap import sites into the arena and pass them out—in addition to posters supporting him that I printed on a poster-printer press I'd bought just for this purpose. The power of our little community behind him was impressive.

When Jesús was in the middle period of his career, he wasn't yet getting top dollar but he had to travel all over the country for bouts. Because he wasn't very well known, a lot of promoters wouldn't pay any additional expenses for a trainer and two corner men. As I was one of the few people from our club able to travel with him, I volunteered to work in his corner. For me that usually meant I was the guy who would bring him a bottle of water and towels when he needed them. Eventually Jesús and I became good friends. One of the greatest honors from him was asking me to work in his corner the night he fought Thailand's Sirimonkol Singmanasak for the WBC super featherweight title. By that point he was getting paid enough that he didn't need me, but he wanted me there. Me, Richard Lord, and his cut man. I've been to a lot of interesting places on my adventures, but working in the corner during a world championship fight ranks very high on that list. Standing in the ring with a good friend when they announced his name and lifted his arm to proclaim him the new super featherweight champion of the world is close to indescribable. Later I was with him when he lost that championship, so I really was there for both the way up and the way down.

In the gym I would often do mitts with Jesús Chávez. After hitting those gloves for several rounds, I was spent, exhausted. At those times I was especially nonthreatening. Jesús would put the mitts down, clasp his hands behind his back, stick his head forward, and challenge me to hit him as he danced to avoid my punches. I tried, but I couldn't. Even when I was fresh, he was far too quick. As his confidence grew he wouldn't wait until I was tired, he'd dare me, saying, "Hit me in the head or the body, anywhere you would like." Occasionally I might get a glancing blow off his shoulder when he ducked, but in more than a decade of working out with him, nothing I ever hit him with made any impact.

Except once. And it remains the single greatest moment of my entire pugilistic career. As usual, I was punching, he was ducking, and I was missing. But then something outside the ring distracted him for only an instant. As I watched his attention being diverted I knew, I absolutely knew, that this was my moment. I hit him with the strongest uppercut I was capable of throwing. I hit him right in the solar plexus . . . and I dropped him. He went down to his knees using one arm for balance and sat there. Then he got up and said, "That is enough for today."

"I won!" I said. "For me, that's a knockout. I win today." It was quickly forgotten, forgotten except for the poster I had printed that still hangs on the wall at Richard Lord's gym proclaiming, "On (this date) Richard 'The Weak Link' Garriott KO'd Jesús 'El Matador' Chávez in the 4th round."

5 CREATE

Minding My Own Business

Far more quickly than anyone imagined, computer gaming became a huge business, and for me there was a steep learning curve from creator and designer to businessman.

When I was making the first *Ultima*s there wasn't much of a business to run. I was self-employed and followed my own schedule. I wasn't responsible for anything except getting the next game done. My job was to be creative, so if I wasn't feeling inspired at the moment I just wouldn't work. Early on, that was perfectly acceptable because no one had any specific expectations of me. The amount of time I spent making a game was measured in weeks or months, while today it requires hundreds of people working thousands of hours.

But after my brother Robert, programmer Charles "Chuckles" Bueche, and I founded Origin in 1983 everything changed. All of a sudden larger amounts were being spent, our own money, and for the first time there was a conflict between the creative process and the business process.

My brother and I had very different motivations. For Robert, it was all about shareholder return and return on capital. His objective was to build a strong company quickly, both to maximize the opportunity and to protect ourselves from competition. But I wanted to work at my own pace. I really was working by myself; as far as I was concerned, the company existed to publish the games that I finished when I finished them. I expected Robert to provide the tools and resources I needed to do that.

It didn't quite work out that way. While I was happy working my own hours, the people who answered the phones and manufactured the games and sold them were working a regular eight-hour day. Too often I wasn't there to give them the answers to questions or the guidance they needed. My brother called me on it, but I wasn't interested in listening to him. It became a very serious issue when we began hiring people to work with me on projects; everyone else was operating in the leader's shadow.

The leader's habits become everyone's habits. If the leader comes in late and departs early, that becomes the way the rest of the company operates. I learned that I couldn't demand more of other people than I did of myself unless I wanted to make my employees unhappy. And the real needs of a business began to creep into what had previously been a purely creative process. Suddenly there was an artist and a sound engineer and maybe another programmer working on my game and we needed to spend more than a brief overlap of time together. We actually had to coordinate our work. My brother was trying to build a company and I was making it difficult for him—which certainly wasn't in my own interest.

We struggled to create and meet realistic development schedules. This haphazard world in which games just sort of appeared from time to time had pretty quickly become a highly competitive industry. The industry was changing so fast that if we took too much time publishing a game, the cutting-edge technology would already have moved beyond it. So to meet the necessary schedule, my teammates, many of whom had been happy working casually, had to make adjustments. It wasn't easy for them or for me. My brother was always hypercritical of me for being the last person to arrive in the office and often the first to leave. If you asked him to describe my work habits for the first five or ten years we worked together, he would say something like: "Richard would roll into the office somewhere between noon and two in the afternoon. He'd play around in his office with his rubber band guns and coding experiments in his mind that had little to do with actually making a game. He might create some code or something else for a few hours. Then come dark he'd leave and we wouldn't see him again until the next day, when he'd roll into the office somewhere between noon . . ."

In my defense, living an unstructured life had worked very well for me. This was my most productive time creatively and economically so I didn't understand why I needed to change. As I wasn't being paid a salary, but only earned an income after a game was published and successful, I felt like I was clearly doing enough to create the next best-selling game. The money was rolling in. Even now, looking back, I still believe the games we created in those early years were the best-designed games we've ever done. So I can correlate the two and make a strong argument that coming in at noon and leaving whenever I wanted to was the best way for me to operate. I've always believed that creativity can't be produced on demand. Sometimes the spark is there and sometimes it's not. I might have been sitting in my office with my feet up on my desk, discussing philosophy with three other developers, or reading a book about ancient languages, but I believed those little experiments were essential to my creative process.

Robert and I are very different people, but there is simply no way Origin could have gone forward without both our unique contributions. In those days we argued often and we argued passionately; both of us were incredibly confident in our own beliefs about the right way to proceed to build a company. And in retrospect, Robert was right about some things and I was right about others. The tension finally came to a memorable head in what has become known as the Battle of the Number 2 Pencil.

The Battle of the Number 2 Pencil

Neither one of us remembers the specific issue that led to this fight, but somehow it came down to the ownership of a number 2 pencil. I started to leave Robert's office with the pencil; he told me to leave it there because it was his office and therefore his pencil. I insisted on taking it with me because I felt it was my pencil. We started shouting at each other and the argument quickly escalated into a shoving match. All of our frustrations suddenly burst. The next thing I knew we were literally wrestling across the tabletops. Then, incredibly, the pencil broke in half.

We stopped, incredulous. We had broken our pencil! When we finally opened the door, all of our employees were standing in the hallway wondering if this epic struggle meant the end of the company. But we started laughing instead. I began to accept the fact that living my life my way was making it much harder for everybody else to do their jobs. We never again had a serious fight.

The compromise Robert and I reached was to establish core hours, a period of time during the day when everybody had to be at work. The morning people could come on time but they had to stay till a certain hour, while the night owls would have to arrive a little earlier than they usually did. For me, at least, the creative process really does need a relaxed, unimpeded period of time. But I don't believe that it hurts the creative process to come into the office in the morning. In fact, I think it undersells the creative process to believe that you have to reject any form of discipline to maximize creativity. I now believe it's possible to be both disciplined and creative.

We definitely made a lot of mistakes in those early days as Robert and I learned how to run a business in an industry that was changing drastically every month. Some of those mistakes came perilously close to forcing us to close down the company. My sister-in-law Marcy once coined a phrase that accurately describes members of the Garriott family, myself included: "Occasionally wrong, but never in doubt," and it was that sometimes misplaced confidence that came very close to bankrupting Origin.

After writing twenty-eight versions of my first computer game on my high school classroom teletype, I began creating my games on the glorious Apple II. I wrote *Ultima*s *I* to *IV* on the various versions of that machine, and the introduction of the Mac in 1984 reinforced my belief that Apple was the future of computers. Apple had always been our lead market for sales; we would write the codes for the Apple then store those codes on a floppy disc until we could find a way to manually read into other machines, including the Commodore 64, the Amiga, and all the other emerging brands. Then we would have to tailor it to make it work on each machine. Every machine stored information differently. It was a fairly difficult process, but we sort of figured out how to get it done. So when we began working on the next game in the *Ultima* series, I saw no reason to change strategies for the latest platform that had been introduced, the IBM PC.

As far as I was concerned, there was no comparison between the Apple II and the IBM PC. The IBM was a big, unattractive rectangular box; instead of a true keyboard it had calculator keys in the "chiclet" style. It had no obvious advantages; it didn't have substantially more memory and its DOS operating system was strange and difficult to use. To me, at least, by every objective measure it was either similar to or a little worse than the Apple.

We had to make a vital decision about how to proceed with our product development. At the time we had eight different games in development and eventually would have to port each one to the numerous platforms, so we looked at the potential of this new platform very carefully. I concluded that the IBM PC would never pose a threat to Apple's dominance of the market. I figured IBM would pick up some market share from many of the smaller companies, but I was supremely confident it would never compete with Apple. So we continued developing our products for the Apple II, knowing we would later translate them to the PC and other machines. I wasn't the only voice in that decision, but I was clearly one of the loudest.

It would have been almost impossible to be more wrong. That was one of my first big lessons in: "What I think is not necessarily right—and perhaps not what everybody else thinks." I realized that if we intended to stay in business, I had better pay more attention to understanding public opinion and its influence in the marketplace rather than relying on my own instincts. I had to accept the fact that there was no way I was going to be able to predict the outcome of these market shifts; instead of basing my decisions on what people ought to do, I needed to think more about what people might actually do.

In the gaming industry at that time, there was no such thing as market research. We didn't have the slightest idea who was buying our games. But it became obvious almost immediately that people were making buying decisions using very different criteria from me, for it turned out that many people and companies had been buying IBM machines for years. IBM computers worked well, the company supported them well, and when necessary they could be fixed or replaced. Consumers trusted the name. Apple was perceived as a young, cool start-up that had come out of somebody's garage; it looked cool and kids liked it, but IBM had a well-established, well-earned reputation. Additionally, competing manufacturers cloned the IBM system and began offering PCs that were considerably cheaper than Apple's.

In less than six months, the IBM PC became the dominant

machine in the market; everything else, including the Apple II, was suddenly irrelevant. So due to my mistake, Origin was caught developing products for a market that was rapidly disappearing. Those half-completed games were essentially valueless, forcing us to start from scratch without any employees who knew how to program on the IBM PC; it used a completely different language that was not easy to learn. We would have to retrain our existing staff and/or hire new people, and would have to delay the delivery of *Ultima V* and our other games then in development for at least six months. Robert calculated it would cost at least $2 million to finish the games, which was almost $2 million more than we had in the bank at the time.

When you make a mistake like this, you can either take your loss or you can double down. Robert and I discussed closing the company and walking away with the several million dollars we had in our personal bank accounts. The alternative was to reinvest everything the company had, everything the company was projected to earn from domestic and foreign sales, and everything that Robert and I could borrow personally. It was a very difficult decision to make. The odds of success were relatively slim. In the video game industry about 90 percent of games fail. The logical decision was to close the company.

Robert and I decided to go all in, and cosigned million-dollar loans. It probably wasn't the sensible business decision. It's difficult to explain why, but it seemed like the right thing to do. Most of Robert's net worth was in company stock, so in many ways he was betting virtual money. If the company went out of business, that stock would be valueless anyway. But I had previously sold stock to build my dream house, a house that cost just about every penny I had earned in the industry. Instead of using that cash to pay for the house, I invested in the company and took a mortgage on the house. If we failed, not only would I lose the house, but my brother and I would lose the company and be millions of dollars in debt. We would be left with less than nothing. But we bet on our capabilities. The race was on to get my next *Ultima* game out with acceptable quality before we ran out of cash.

Every month Robert and I would sit down and look at the numbers compared to our progress, and every month the situation looked more ominous. In family lore we refer to this as "the year of Richard's fetal position." When we reviewed the financials, I found it so difficult to listen that I would sit in a chair in a corner in the fetal position with my legs curled up under me. As Robert went through the numbers, I would sit there rocking in horror and disbelief. After all the success we'd enjoyed, it was extraordinarily difficult to deal with the possibility of failure—especially given that my mistake had brought it on.

I barely slept at all, and when I did sleep the nightmarish scenario of what I could have, should have done played over and over in my mind. I would often stay up most of the night playing the very games that I believed would be my downfall.

Desperation can be an extremely productive motivator. It is a fact that no game ever comes out on time. Ever. That's just the way it is. But we had only enough money to keep the doors open till the original ship date. If this game wasn't published on time, and with the highest possible quality, we were through. Since we had never managed to do that previously, there was good reason for that fetal position. We worked every possible minute for months, while at the same time resisting the urge to cut quality just to get it done. It was an incredible balancing act: we kept everything that was necessary, but if something wasn't absolutely essential and would cost us more time, it got cut out. As a result, this was our cleanest execution of a game.

Somehow we managed to ship it on time. For two or three days we were elated, ecstatic—we'd done it! We'd done it! And then the phone rang in our customer service department—and within a couple of hours we were despondent, certain that this time we really were going out of business.

Some of the initial players had gotten the game and very excitedly tried to install it—and it did not work. One of our programmers then tried to play the game on a floppy disc, and discovered the problem: we had all been testing our work with the newfangled hard-disc drive and no one in the entire building had thought to test it on a floppy-disc drive. It took about five floppy discs to complete the game, which was so time consuming and unnecessary that none of us had ever played the game on a disc. In-house we worked on hard drives, then copied it to a floppy and shipped them. It turned out that if you tried to play the entire game on floppy-disc drives and did not have a hard drive, it failed. That wasn't true only for the person who called or for our programmer, it was true for every person on earth who did not install this to a hard drive. And at that point less than 10 percent of consumers owned a hard drive.

By the time we discovered that *Ultima V* didn't work on a floppy disc, we had shipped more than a hundred thousand boxes. The old saying we didn't know whether to laugh or cry didn't apply; there was absolutely nothing funny about this. We had blown it. There was nothing we could do but wait for the other shoe to drop as people who bought the game discovered they couldn't play it. I got back into my fetal position and waited.

But the flood of calls we expected from the stores didn't happen. In fact, the next day we didn't get a single complaint. The day after that we may have gotten one or two, but nothing at all like what we expected. Why weren't people demanding their money back?

Miraculously, that small percentage of people who had hard drives included a high percentage of gamers. Gamers had bought them as soon as they were available because this was the precise population that needed them. We got back about the same number of boxes that would have been returned due to defective discs or similar problems. I managed to unfetal myself. We had been saved.

6

CREATE

A Universal Language

Disaster averted, we got back to work making games, and this time we were able to focus again on the fun stuff. I've always been fascinated by languages. A key question I've long wondered about is, when we finally make contact with other civilizations, how will they communicate with us? It won't be a spoken language we'll recognize; they won't say anything as simple as, "A funny thing happened on the way to earth." And we won't raise our hand in peace and say, "Welcome to earth." It's possible their language will have a symbolic representation—but we might not recognize their symbols. For example, the concept of a stick figure representing a person is great if you're a human, but if everybody on an alien planet looked like a slug that would make less sense. In the movie *Close Encounters* musical tones were the basis of communication, which was almost immediately taken over by computers teaching each other how to communicate. Now please just pause right here for a moment and ask yourself: How would you communicate with an alien? For example, how would you communicate the thought to them, "Please don't blow us up with your cool weapons?"

I've spent a lot of time working on this problem. Crafting believable languages and scripts is essential to creating any realistic fantasy world, since language and writing is the foundation of a society. Tolkien, the greatest reality crafter I've ever discovered, once said his stories grew out of the languages he created. On the very first page of *The Hobbit* there is a map of Lonely Mountain and in the bottom corner, written in a script he named Dwarvish, which is clearly derived from early Celtic runic, is the direction, "Stand by the grey stone when the thrush knocks, and the setting sun with the last light of Durin's Day will shine upon the key-hole." Runic is decipherable because it has enough letters that are recognizable to people who understand the Latin-based English alphabet. As soon as I read that direction, I did what I'm sure many people did—I turned back to the map and realized the words on it were not fake symbols, but all had a meaning I could understand. The number of words and the general number of characters per word matched the English-language poem I'd just read in the book. I suddenly realized these strange symbols must be that same poem. As I stared at it, I realized there were even some letters—for example, the letter *r*, that looked like their modern counterparts. By glancing back and forth between the poem and the map I could slowly work out what each symbol meant as a sound—and sure enough, the entire poem was written in those symbols. The revelation to me that this map was "real" was powerful.

Among the many other languages Tolkien created was Elvish. As hard-core a Tolkien fan as I am, and I am a pretty darn hard-core fan, it's too difficult to learn. That Elvish is too complex for most people to understand remains one of the few disappointments I have ever had in this great work.

Maybe the most important lesson I learned from Tolkien was the importance of building a story on a strong foundation. To make a narrative feel real, and to get players emotionally invested in it, the backstory has to be complete. For example, I'm a huge fan of the Disney theme parks because of their devotion to the purity and depth of the story. The entire time you're standing in line for a ride everything you encounter reinforces the conceit that the world you are about to enter is real.

Only once did Disney disappoint me. The line for its Indiana Jones ride is themed; as you move through it, you encounter old Indiana Jones car parts and see people on a dig outside Cairo, and finally you see blocks of pyramid stones. Written on them are hieroglyphics. And when I looked at them, I was shocked. There were probably ten characters on the top line, but when they needed a second line all they did was turn the first row upside down and stamp it into the pyramid. It was obvious someone had been too lazy to continue the conceit. That emphasized to me what an incredibly important difference the smallest details will make in the creation of a fantasy world.

In school, the Runic script I had learned from Tolkien proved to be very important. It was at least part of the reason I was a B or C student rather than failing many courses. Because no one else could read my Runic, including my teachers, I could write down anything I wanted and keep it in plain view. Before an exam I simply would write all the information I might need, in Runic, on the cover of that subject's folder and when I came into class I would drop the folder down on my desk or on the floor where I could easily see it. On the one hand, I was cheating, and cheating deserves punishment or an apology from the cheater, but on the other hand, it was pretty darn clever cheating. And it probably took more time to write my notes in Runic than it would have to actually sit down and memorize the material.

I've created numerous languages for my games. While I have

delegated many aspects of world building to other members of my team, I've kept the creation of virtual languages all for myself, mostly because I am the only one who takes it so seriously. Successfully creating a powerful and unique language requires both the proper motivation and an extreme devotion to researching language development.

The first "language," or more properly "script," I created for *Ultima* was simple: It was just a version of Runic, the same Runic of Tolkien's books. Ancient Druid runes were the basis of the Roman language, which was the basis for the English language. So the symbols and shapes are similar to our language. The Runic letter *b* looks just like a *b* but with angles instead of curves. Roughly a quarter of Runic letters have similar-looking English equivalents, though because they had to be chiseled into stone, Runic letters generally had a more angular shape. The pronunciation of the letters is also vaguely recognizable. Since Runic was the predecessor of the Roman letters we now use, Tolkien did not try to make pronunciations different, only appearances—but these Runic letters still represent Roman letters. So my Runic-based language was also a simple English substitution with twenty-six letters, though it included a few digraphs, like *ea, st*, and *th*. Players could easily substitute letters and read. It was relatively simple to identify Buccaneers Den on a map, for example. The *b* looked like a *b,* the *u* was an upside-down *u,* the *cc* was identifiable as *kk*. With just a little effort a player could figure it out—as long as they understood English.

However, if a player did not understand English, Runic was shockingly difficult. And that's what happened when *Ultima* began selling around the world. Japanese players had to convert the Runic into English letters and determine that it spelled Buccaneer and then figure out what the heck a buccaneer was. Although the games were sometimes nominally translated into foreign languages, the maps were left in Runic. While the first few *Ultimas* became cult hits in non-English-speaking territories, their impact was limited by this language barrier. The texture of *Ultima,* for example, required the player to become completely enveloped in Britannia and that proved very, very difficult without fully understanding the language. The American games that made a ton of money abroad were first-person shooters that did not require translation.

So I reconsidered my entire approach to language creation. My original goal in using Runic had been to add a level of mystery and secrecy to the society I was trying to create in *Ultima,* while also making it as easy to decipher as possible. The fact that it was a script that could be used to understand an actual message hidden on the map—and not just a collection of meaningless symbols—further reinforced the realism of this fantasy world. When I realized its limitations, I began looking at other languages. I wanted to understand the underlying concepts that could help me create an easy-to-read universal language. As I discovered, that was no easy feat. It had been tried before: Esperanto was created to be a universal language. But while Esperanto makes grammatical sense, there is no intuitive way to know the meaning of a word without being told what the word means. In that sense it is no different from any other existing language.

I wanted to create an entirely new language for the game *Tabula Rasa,* a language that could be learned from a few fundamental visual clues. I wanted to do it with pictograms, hoping that once a person recognized the meaning of the first few symbols, they could decipher most of the others without being taught. I wanted all players across the world to enjoy the same aha! moment I had when I realized the wonderful game Tolkien was playing.

The story of the language was simple: Long ago, when aliens had visited this planet, they had found only lower life forms incapable of understanding their advanced technology. So they'd left behind a message, written in a language that would be understood only after society had advanced to a higher level. The message simply announced that they had been here countless years earlier.

But to make this believable I needed to create a language that conceivably could have been used by an advanced civilization. I began by looking at a variety of other languages. Most gamers are familiar with *Star Trek,* so I looked at Klingon. A Klingon dictionary has been published and people have attempted to turn it into a complete language. But I'd seen every episode of *Star Trek* and recognized that Klingon was nearly impossible for a normal person to speak. It's more difficult to learn than almost any other known language, in part due to very challenging pronunciation. There is some logic to it, but much less so than languages that have developed through time and use.

I needed a language as easy to comprehend as Runic, and one that was equally accessible to people who spoke English, Japanese, Russian, Arabic, or any other language. It had to be robust enough to allow the expression of any thought, ranging from "Is there a god?" to "Pass the ketchup." Rather than letters it had to be some form of ideograms that could easily be communicated from one person to another. So I knew what I needed to do; all I had to do was create it.

Go ahead and try it. Think of one symbol that any intelligent creature, no matter what language they speak, no matter what planet they evolved on, will be able to understand. One.

That's when I began my research in earnest. Most people think of Egyptian hieroglyphics as the most common symbolic language, so I began trying to understand it. But it turns out that hieroglyphics are not really ideogramic. The pictures represent sounds, not objects, which is why no one was able to interpret them for thousands of years: unless you knew how to speak ancient Egyptian, the sound signs had no meaning. So hieroglyphics weren't at all helpful to me.

Then I began considering the international signs for the

Olympics; it's an event attended by people from just about every country on earth, so the signs have to make sense in any language. And, in fact, some of them are universal. Essentially these signs allow any individual on earth, no matter what language they speak, to find a bathroom and other necessities—but it's a very limited language and it's not possible to communicate complex ideas. I also studied hobo signs telling other hobos, Don't run through this guy's yard because he has a shotgun, or That house gives great handouts. But these signs aren't pictographic and they vary from region to region.

The first language that felt like it was on the right track was ancient Chinese. Modern Chinese is a very simplified version of ancient Chinese pictograms. When the ancient Chinese drew a picture of a house, it meant, This is a house. That was a start, but it still was limited in scope and sometimes the pictures seemed arbitrary to me. And it was hard to express complex ideas.

It was then I discovered the key to the creation of my next language—a book entitled *Semantography/Blissymbolics* by Charles Bliss. Blissymbolics is a universal pictographic writing system that enables people to communicate with mentally handicapped children. Some of it made perfect sense, like a rectangle representing a sheet of paper. But a lot of it was unnecessarily complicated. It had a whole system for modifying symbols, turning them into action verbs or adjectives, or indicating whether they were singular or plural. I don't even know my adjectives from my adverbs very well, so I didn't think mentally handicapped people would find these details helpful either. And truthfully, I didn't care about adjectives or adverbs. It didn't have to be a big house and the person didn't have to run quickly; it was a house and the person ran. I wasn't interested in tenses or modifiers, I just wanted to communicate an idea directly. What I needed was a base vocabulary: a series of core shapes and a means for them to interact with each other so that a few symbols could convey as much meaning as possible.

I followed my usual creative process and started collecting ideas without trying to categorize or assign a value to them. I still have a coffee cup crammed full of little slips of graph paper on which I've written symbols that successfully communicate basic concepts.

I ended up with a series of symbols that were simple to modify. For example,

means long. An hourglass indicates time. Placing a dot to the left of the hourglass means the past; a dot on the right means the future. These two symbols communicate the classic opening of a story; the first column (below) reads: "Long ago our civilization grew on a planet which was a far travel into the stars."

My greatest challenge was how to communicate personal pronouns. The basic symbol I used consisted of two connected figures, meaning us, we, them, or our rather than I. I decided that a single quote mark above the figure means first person, I or me, two marks means second person, we or us, and third person, three marks, is them. A house obviously is a house or a shelter, while two houses represent a community or a civilization. A very small person lying down is a baby, and a standing small person is a child. It requires only a single page of symbols to decipher the entire language, and once those symbols are understood, there is nothing that wouldn't make sense. It's not that hard to read.

I tried to create a variety of base symbols that could be used in multiple ways to say a group of related words. I decided early in the process that I wasn't going to be concerned with grammatical structure. The purpose of my language was to communicate an idea; I didn't care if it was communicated good, or well, or correctly.

For games this concept worked very well, but the language couldn't be spoken easily. There was also no way to express proper names. I realized I needed a third language to address some of these shortcomings, and I have been working on it for almost two decades. A rough version of it was presented as "Gargish" in *Ultima VI,* but it continues to be a work in progress.

Gargish is a phonetic script with a symbolic component. As I was working on it, I extensively researched spoken languages. I learned that all the sounds that make up words can be broken down into a few foundational mouth positions, and more specifically, four distinct uses of a similar mouth position. For example the sound *b* is voiced as ba. This is a bilabial sound, meaning we use both of our lips to make it. Try it. The *p* sound, pa, is made in a very similar way, although one is voiced with your vocal cords and one is not. These are plosive sounds, meaning the sound starts with a burst of pressure out of your mouth as it opens.

The *f* sound, fa, and the *v* sound, va, are pronounced with your mouth in the same position as ba and pa, but they are not plosive. While there is a "start" to the *b* and *p* sounds, *f* and *v* are more continuous, no bursting of sound at the beginning. So that's four variations on one mouth position: Two of them are plosive and two of them are long. One of each of those is voiced and one is not. So, *b, p, v, f* are four sounds in a two-by-two matrix.

It's possible to go through all the various sounds that make up our language and group them—just as I did when determining the three virtues for *Ultima IV*—it turns out you need only five symbols to represent all the mouth positions. By adding a few indicators for plosive and voiced, suddenly you've got an entire language, which can be spoken by anyone who knows the rules. I also added cartouches, ovals that first appeared in Egyptian hieroglyphics and indicate that a set of symbols represents a proper name.

My languages have helped make my worlds as realistic as possible. In addition to appearing on cloth maps, and being used within games to convey especially secretive clues, I have also used them to carve stories of the past history of this civilization into buildings and obelisks, just as the Greeks and Aztecs did.

In a few instances, as with Elvish or Klingon, players of my games have used these languages to communicate outside the game. When players learned that I would personally respond to letters written to me in *Ultima* Runic, I began receiving piles of mail. In fact, letters written in that language were guaranteed to get to me because nobody else in the office could read them. One of them was written by a person I would call the ultimate *Ultima* fan, a man named Joseph Toschlog. He had decorated his family room with *Ultima* memorabilia, including all the maps and many artifacts from the games. Other than that, his personal passion is cooking. He went through all of the *Ultima*s and extracted each comment a character made about a food product that existed in Britannia and assembled them in a Britannia cookbook. For recipes that didn't exist, he imagined what they might have been and created them. When we were running a Kickstarter campaign for our new game, *Shroud of the Avatar,* he joined as a backer—and also sent the team a large box of food from his Britannia cookbook. He enclosed two letters, one congratulating us in English, and another, to Lord British, in perfect Runic. It was an invitation to a feast he was hosting to celebrate the founding of New Britannia. There is something very satisfying about reading an invitation written in a language you have created.

While we could not attend his event, we did something better—we hired him. He worked for us for about two years, helping in our crowd-funding campaign. And during that period we held several feasts in which Joseph played a central role, twice serving as host and chef.

There was at least one other time I used the language of our games to communicate directly—and only—with our community. It was the one place where the two worlds in which I spend much of my life collided: When I made my trip into space, I knew that only two minutes into the launch process there would be an onboard camera inside my Soyuz spaceship that would be broadcasting to earth. I wanted to send a message to players of my games, especially *Tabula Rasa,* letting them know they too were a part of my journey. So, on the back side of the launch manual I would be using, I added a message that faced the camera, written out in my Logos language. No one on earth except players of *Tabula Rasa* could have read that message! Let's see if YOU can decipher it!

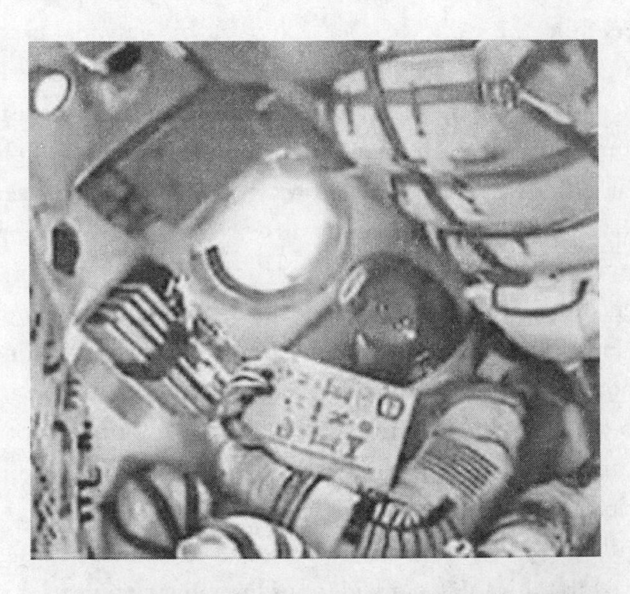

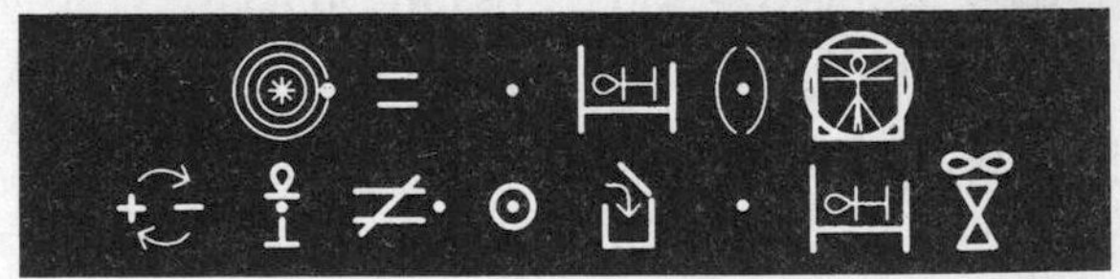

7 CREATE

IV-Play, or The New Rules of Engagement

Thou hast sworn to do thy Lord's bidding in all. He covets a piece of land and orders the owner removed. Dost thou: Keep thine oath and unfairly evict the landowner or Refuse to act, thus being disgraced?

Ultima III: Exodus was the first game Origin published. Before that I sat by myself and wrote a game, and when it was done put it in a box and sent it away. That was it. I never heard another word about it. If the publishers received mail, whether praise or complaints, I never saw it, and the Internet did not yet exist. Other than sales figures and a few formal reviews in computer gaming magazines, I really had no idea if or why people liked my games. Now that computer gaming has become a multibillion-dollar industry, companies pore over data and use focus groups to create games directly serving their audience, but back then we got no feedback. So I was more than a bit surprised when Origin began getting fan mail and Robert and I found out that players weren't playing my games at all like I had intended.

I had assumed people played the game to be the hero, the

noblest person in the kingdom. I was wrong. Most people were doing what we in the gaming industry call "min-maxing" the game: finding the most direct path to power without the slightest regard for morality. In almost every role-playing game, the general goal is to defeat the main antagonist. From reading fan mail I learned that because there was no real motivation to care about the bad guy, there was no motivation to act morally or virtuously. Instead, players were spending the minimum time and effort to reach the maximum reward, victory. They didn't care very much about the story, only about advancing.

If there were shortcuts, they took them. If they went into a store inside the game to buy a sword, for example, and discovered that by turning off the computer after receiving the sword but before paying for it, they would have the sword for free when they resumed playing, they would turn off the game. That was a way to min-max the game. And there were many other tricks like this for becoming wealthy. Players were maxing the system, killing, pillaging, and plundering in their quest to become powerful enough to kill the bad guy. If they could get wealthy more quickly by killing innocent villagers in town rather than the horrible monsters outside town, that's exactly what they would do.

Equally surprising to me was that they didn't care about my bad guy because he wasn't bad enough. As in most other games, my bad guy was just waiting there at the end, the last character players had to defeat to win the game. So players would min-max their way to the top in a morally ambiguous way and finally kill the bad guy for no reason other than that was what was required to win the game. There was very little emotional involvement; in fact, as far as we could tell, the greatest emotional involvement people enjoyed came as they described their massacre of all my villagers and their attempts to kill Lord British.

When I wrote *Ultima I,* I wanted the game to be something more than players wandering around 3-D dungeons killing things. But while we had successfully added some pretty cool visual elements and interactions for the era, these fan letters made it clear the game hadn't progressed as I'd hoped. Two things became obvious to me: First, if there is going to be a bad guy, he needs to actually be bad in a way far beyond what the player is told in the introduction. He has to be an active presence. If I am a player seeking advice from confidants throughout the world, the bad guy should be looking for these people and killing them. He should be responding to my actions and making my journey far more difficult. Rather than simply existing and waiting for me, he should be trying to do terrible things to me and the people I care about as I progress toward a final confrontation.

Second, I was surprised that the people playing these games did not play with the moral compass they followed in the real world. Presumably, they were law-abiding, kindhearted people in their day-to-day lives, but in this fantasy world they enjoyed being someone else. That was understandable, as my games didn't respond directly to their behavior; there were no rewards for being a moral person and no particular penalties for being an evil wizard.

Five years into a very successful career I had an important decision to make. My other games had borrowed liberally from existing fantasy stories: None of them were particularly original, other than the fact that they were being told in a computer game format. I had adapted the basic plot structure from *The Lord of the Rings*. Players were able to move back and forth in time as did the characters in *Time Bandits*. I had borrowed elements from my own D&D games and from Narnia. *Ultima II* became the first game ever sold in a box because I wanted to include a cloth map like that in *Time Bandits,* and wanted to disguise the instructions as an elaborate history of my world.

All of these stories had inspired me. But if I was ever going to compete with the writers I most respected, I knew I couldn't continue stealing from D&D, *The Lord of the Rings,* Narnia, *Star Wars,* and *Time Bandits.* Games, I realized, had the potential to be much more than a pleasant diversion; done well, they could be as much an art form as books and movies.

I wanted to write a game about virtue, a game in which the player was judged not only by the fact that they had risen to power, but by the path they had taken to get there and the lessons they had learned along the way. There would not be a traditional bad guy. And rather than the player just stomping through the land wreaking havoc, I wanted to hold a mirror up to the player's behavior. I wanted to force them to be virtuous in order to win the game. Rather than merely killing monsters, I wanted to weave an enriching story that the player unraveled through his or her own behavior. And show them, through their own actions, that maybe they weren't as virtuous as they thought.

I've always been a devout believer that role-playing can be a powerful experience and has the profound ability to teach. I've also believed that the very best stories in any medium allow the audience to reflect upon themselves, their beliefs, and the human condition. They enrich you as a participant, and when you're done make you feel that you've spent your time well, that you've grown through that experience. That's why when I'm telling my own stories I will always stop and ask listeners to relate it in some way to themselves.

In other words, I thought, How cool would it be to incorporate a player's own value system into the game? If I could do that, instead of the player mindlessly plodding along killing monsters and collecting treasure, players would be forced to emotionally invest in the game. Instead of pushing a button, they would have to sit and think and make moral choices.

When I told my brother and our team I wanted to tell a story about the player's spiritual growth through their own actions, the response was pretty much unanimous: Are you nuts? There's no profit in virtue! That's a terrible idea! Richard, it was explained very carefully to me, people don't play games for the beautiful story. They play for the adventure. People want to fight monsters and kill villagers, that's what they're doing to have fun—and you're going to take their candy away from them? That's a recipe for utter disaster. And worse, financial failure! Actually, I responded, my idea was worse than that. In this game I wanted to penalize players for the bad behavior that earned rewards in other games.

Players want to escape reality, I was told, not dive deeply into profound personal issues. But I was determined. It seemed obvious that meaningful contemporary social stories could be told in my medieval setting. Just as I believe *Star Wars* could have been set in a medieval world or *The Lord of the Rings* could easily have been set in a futuristic world, I believed I could tell personal stories about our world in a game. I could make the game meaningful.

I understood that making storytelling the dominant aspect of a game was very risky. When I pointed out that no one had done it before, I was reminded of the reason for that: that's not what players were buying.

But I was certain I could make a storytelling game people wanted to play. Or I thought I could. Or I hoped I could. Okay, honestly I had some serious doubts I could pull it off. I knew I could figure out the technology. The real question was which sorts of behavior to reward.

The Prophet of Truth

What are the most important virtues? Just stop for a minute and put yourself in my position: you want to create a game that is going to judge good behavior. Where do you start? Make a list in your mind of at least the top three, or four, or five virtues or the commandments that you believe are most important and realistic to embrace. And don't forget it.

Many people might start by looking at religious values. I went to church when I was young, and my eldest brother and his wife are quite religious. But it didn't really take with me. I always ran into trouble when an historical doctrine was in conflict with the scientific process. But I do accept that religions are systems of beliefs that attempt to teach positive values. The problem is that they're not all the same virtues. So which religion should one follow? I think much of the Old Testament dogma is no longer relevant. How could I proclaim that people should not eat pork if I believed that law had been made specifically to protect people from diseases common at that time but that were no longer a threat?

I needed a moral code that I felt good about espousing as the Truth (with that capital *T*) and one that was not going to make people think I was (a) lying, (b) promoting a particular religious belief or institution, or (c) making something up that I didn't personally believe. That's a big challenge.

How's your list coming? Let's make it easy: What's first? What is the most important virtue?

Research has always been the foundation of my creativity. I begin any creative endeavor by learning as much as I can about the subject, gathering information before I develop an idea. In this case, I started by reminding myself that this was not an area I'd studied in school. I never studied philosophy; I never studied comparative religion; I hadn't enjoyed history. But if I was going to make a meaningful contribution to this new art form, I had to become interested in things that had never interested me before. History especially: facts and dates didn't compel me, and it wasn't until I began to understand how civilizations were often built on political intrigue that I felt the slightest interest in it. So I began refining my understanding of history, and by association, philosophy. Every game I created grew out of my passion to deeply understand a subject.

It was a journey. I bought a library of books. I mined the works of Khalil Gibran for an understanding of virtue. Jean de La Fontaine's fables became a source of scenes I could set to challenge a player's moral compass. French designer Thierry Mugler's books became an inspiration for timeless fantasy and characterization. When I looked at history I wondered if I should start with the virtuous counterparts of the seven deadly sins? Would the seven heavenly virtues be a good basis for a game? The seven deadly sins are usually described as wrath, greed, sloth, pride, lust, envy, and gluttony. That wouldn't work, I decided, because I don't believe those seven deadly sins are universal, or the only ones that are relevant or even unique.

I wondered about the universal acceptance of the ten commandments. Who could possibly disagree with virtues like do unto others as you would have them do unto you? Then I realized that ten weren't enough, that there are other equally important things that should be included, and that they're redundant.

If the virtues of Christianity weren't sufficient, then what about Buddhism? There certainly are parts of the four noble truths, the path that leads to awakening, that I really like. Among all modern religions, I am perhaps most intrigued by Buddhism. While I do not describe myself as a Buddhist, I agree with their philosophy that the purpose of life is happiness and the way to achieve that is through compassion. I personally would add that if you don't have a life goal of happiness, you probably won't achieve it.

I exhausted my reading list, and finally realized there was no list of suitable virtues written by somebody else. I would have to develop my own.

So I continued to explore philosophy, especially philosophical poetry. I copied snippets of Gibran, believing that if characters in the game were going to tell people to live compassionately, they should do so eloquently. If I was going to quote Gibran, maybe there were other people I should know about. I bought quotation dictionaries and looked up words associated with virtue, like truth, compassion, honesty, valor, and courage, to see what great thinkers had said about them. I wrote the quotes that appealed to me on Post-it notes and stuck them on the wall. I was searching for a distinct set of precepts. But as time passed and I read them over and over, I saw that many concepts were closely related. For instance, kindness and compassion certainly are not the same thing but they are similar; a compassionate person would have to be kind. So I began to organize my quotes into similar topics. That forced me to really examine each virtue: valor and love are kind of distinct from each other, for example, so they each deserve a separate category.

The deeper I searched, the more complex this became. There seemed to be no limit to the subdivisions of subdivisions of subdivisions of energy and forces. So I simplified: If I was going to create a moral code or philosophy that I wanted players to accept, it had to be logical and fairly basic. I finally settled on three cardinal virtues.

This was not exactly like reaching the peak of a mountain and being handed an inscribed tablet. I was always aware of the fact that this was fiction. I watched it unfold—for a long time it was like watching static on television—but gradually it came into focus for me. I came to the conclusion that almost every quote I liked and put up on my wall had been derived from three unique motivational principles: truth, love, and courage. Without too much difficulty I could describe almost every other virtue represented on a Post-it note as a combination of those principles. And vices could be described as the absence of one or more of those three principles. I could account for the entire universe of moral philosophy in the success of those three archetypical principles: truth, love, courage. As I added nuance, I understood that these three principles can be combined in some way to make eight virtues: honesty, compassion, valor, justice, sacrifice, honor, spirituality, and humility. Truth and love, for example, are justice. Love and courage are honor.

These principles and virtues became the foundation of my game design. I could build three large castles, one to truth, one to love, and one to courage, and eight cities, each founded on the ideal of one of my virtues. I made eight dungeons, each one the opposite of one of those virtues. I created characters that would be the leaders of these schools of thought.

Using the three principles and eight virtues it was easy to make a pattern for players to understand and explore that would feel real to them. I was confident that not only could I make it appear to the player that this was the Truth (again, with a capital *T*) in the Stephen Colbert sense of truthiness, I knew that mechanically I could also make it work in the game. I could teach it in a structured way that seemed to have a rationale behind it that satisfied my three original conditions: (a) had a firm basis in knowledge, (b) did not promote any particular religion or philosophy, and (c) accurately reflected my own beliefs.

Throughout the development process many people in the company continued to think I was nuts. The *Ultima* formula was proven and profitable, and many saw no reason to change it.

I believe *Ultima IV: Quest of the Avatar* was the first video game in history to focus on morality rather than killing. Some reviewers thought it was intended as a response to those people who had branded video games as a perversion that encouraged antisocial behavior and even satanic worship. It wasn't, not directly. It was my attempt to further develop an emerging art form.

This was my first game in which the player had a specific identity. Since the game would judge behavior, it was important that the player thought of their character in the game as their real self, not an evil alter ego. I wanted the player to identify emotionally with their character rather than being a puppeteer of a character the game developer had created. When you played your character in my world, you might be physically transformed into a hulk, but spiritually it was your soul in that body. You had gone through a portal into my world and chosen a different body, but in every important emotional and decision-making way it was still you.

I wanted people playing my games to believe their character accurately reflected their own feelings and beliefs. At that time the concept of a gaming "avatar" did not yet exist. I learned about avatars while doing research on philosophy and virtue; Hindus use the word to describe the physical manifestation of an earthly belief or philosophy. When Hindu gods manifest on earth, they take the form of the god's avatar. I thought, This is the player's manifestation in my world, the world of Britannia. So your character in the game was "you" manifested from earth to Britannia, and "you" were trying to become the avatar of virtue. For a time I actually owned the trademark on the word, but it became so common so quickly that it was impossible to protect it, and there really was no reason to do so.

A manual that came with the game told the story of Britannia as it had unfolded in the earlier *Ultima*s. But this was a new Britannia, a world without elves, dwarves, or "bobbits," a world in which, as Lord British explains, people are searching for "a new standard, a new vision of life." To guide them from "the age of darkness into the age of light" they needed a role model—and the player's quest was to embody each of the eight virtues and become that person, the Avatar of Virtue.

Each of the eight cities of Britannia represents one of the virtues. The player has to visit each city and master that virtue. In each city the player is sent off on a quest of self-exploration, to see how much or how little the player is in tune with that virtue. Tests throughout the game demonstrate a player's honesty or dishonesty, their compassion or cruelty. A player can show compassion, for example, by donating gold to beggars or letting non-evil characters such as snakes or rats or spiders live. He can demonstrate honesty by properly paying a blind shopkeeper—or dishonesty by stealing from the shop. Or win bravery points by the way he responds to being attacked by an evil monster: if he runs away despite being stronger than the monster, he's penalized with a cowardice point. The important thing was that the game was silently watching the player's behavior—and remembering it. If you were a scheming, thieving bastard and took something from a shop, maybe you would be able to get away with it at that moment—but the game would remember it and would penalize you for it later.

Players quickly figured out that the game was watching their behavior—but what they didn't know was when. It was impossible to test every possible choice a player made, but I knew that by testing a handful of them—and not letting the player know which handful I could test—players would believe the game was testing them a lot more often than we actually were capable of doing, and would take the appropriate "virtuous" actions.

Here's a test for you, dear reader: You are deep in a dungeon and poor defenseless children are trapped in cells around the perimeter of the room. It looks like their fate is not going to be good. They are being held by truly evil beings. You have come through some terribly torturous places to get here. There is a lever in the center of the room. If you throw the lever, the children in the cells will be freed. In your pursuit of enlightenment, would you throw that lever?

What players didn't know was that we didn't have the technical capability to test behavior in the dungeon. In that situation, no matter what the player did, it had no effect on their progress.

What did you do? Did you choose to release those children? Did you choose to save their lives? Most players do and take satisfaction in that action.

Too bad. In fact, those weren't good children at all, they were monsters that looked like children and if you released them, they attacked you and tried to kill you, so to get out of that dungeon you had to take them out first! I loved that scene; I put it near the end of the game so the player would assume his behavior had a real impact on winning.

What I had never expected was that one of our beta testers, people who play the game before it's released for sale, was outraged by that scene. He wrote a letter to Robert, the president of the company, telling him angrily that he refused to work for a company that so obviously supported child abuse. "Richard has created a scene where I release these children and the only way I can get out of the room is by killing them. That's outrageous. I demand you remove that scene from the game."

My brother practically broke another #2 pencil in despair. "What the hell have you done this time?" he demanded. "Why did you put this horrific scene in your game?"

In fact, I was thrilled I'd been able to generate that type of passionate reaction. This tester clearly identified emotionally with his character. I explained to my brother that the tester didn't have to kill those children: he could have chosen not to pull the lever. Or he could have used magic to put them to sleep. Or he could have charmed them and made them run away. Or punched them in the nose, causing them to run off. And by the way, if this isn't a test, kill them: they're not children, they're monsters. They are disguised as children.

My brother listened calmly and then told me to take it out of the game. Immediately. If our employee was having this reaction, we were going to have serious problems when we published the game, he insisted. *Good Housekeeping* would come out against us, and all the other organizations critical of the gaming industry would also come out against us. We would be in real trouble.

Was teaching the lessons of my three principles more important to me than profit? This challenge to my own principles seemed similar to what I wanted people in the game to experience. I wanted players to stand up for their beliefs, so clearly in the real world I had no choice but to stand firm. Forget it, I told my brother. My reaction forced him to take the ultimate extreme position: he told our parents! When this had happened in the past, usually our dad took Robert's side and our mother took my side. But in this case even she disagreed with me. "Richard," she said, "it's not worth it. Why is it so important to you to keep this one room out of the thousands of rooms in your game? What's the harm in taking it out? Just this one time."

I can be stubborn when I'm convinced I'm right. "Forget it, I'm not taking it out of the game. You can publish this game or not, but I'm not taking it out." My stand wasn't about pride or ego, it was about my sincere belief that the events portrayed in the game would properly provoke an emotional response in which the player would be forced to conduct a gut check against their actual personal morality. If I expected people to be true to their beliefs, how could I so easily disavow my own? I couldn't, and that knowledge reinforced my determination that this room had to stay. I threatened to stomp my feet, hold my breath, and go to my room! I was serious to the point of being unreasonable.

The game was published with this room in it. We didn't get a single complaint letter.

But there is at least one person who has noticed. He put together a really funny YouTube compilation of that point in all my games where the player may have to kill children. "What is it with you, Richard Garriott?" he demands, not knowing the whole history. "Why do you keep having to kill children in every game?"

When *Quest of the Avatar* was published in 1985 it was very successful, with sales surpassing all previous *Ultima*s. It also has had a lasting impact on the gaming world, commonly appearing on lists of the ten most important video and computer games ever published. It is often credited with demonstrating that games can be a force of social good.

I know it affected people. One couple who came up to me at an event told me, "Look, my wife and I met about the time we began playing your games. We don't want you to think we're freaks because we clearly understand that your games are fiction, but we actually think the message inside the game is exceptionally positive and, in fact, the rules of life that you put forth are logical, consistent, complete, and something we really admire. So when we got married, we had your three principles, truth, love, and courage, inscribed on the inside of our wedding bands."

I also remember the letter I got from a woman who had bought a copy of *Ultima IV* for her daughter. It was the most beautiful thank-you letter I've ever seen. The woman hadn't wanted to buy the game. She was skeptical about the violence in many video games, and believed them to be a waste of time. Her daughter already had trouble in her life; she had been caught shoplifting more than once and was struggling in school. She was hanging with the wrong crowd and her mother didn't think she needed any other negative influences. So when her daughter started to play the game she decided to watch her, and if she thought it was too negative she was going to turn it off. But she was stunned when she saw it was just the opposite: not only did she see a game that she described as "truly enriching," but it also changed her daughter's behavior. She knew for a fact, she wrote, that her daughter had grown as an individual from playing the game. She went from being someone who didn't reflect on her impact on those around her and the world at large to understanding the impression she was making on others. The way she played through the game forced her to hold a mirror up to her own behavior and led to change.

I continue to get letters from parents who have played alongside their children and tell me they have seen a fundamental change take place, perhaps the best validation I can imagine of the moral system I developed. In fact, we used to joke that if we really wanted to, we probably could make a pretty darn good religion. As we all know, there is a lot of money to be made in the business of selling belief systems. The reason we didn't is because I consider it immoral. The entire concept of creating a fictional philosophy based on positive virtues in order to gain power and notoriety, and make a lot of money, is reprehensible to me. I'm careful to never tell anyone what they should or should not believe, because I can't answer that question.

After almost three decades the principles and virtues that form the basis of *Ultima IV* are still talked about and accepted. The game is credited with changing the video game industry, but it was really just a beginning.

8 CREATE

Making Enemies

If I truly wanted to give people a personal emotional experience, I had to give them a world with which they could truly interact. A world as real as possible, one that might exist somewhere over the horizon. A world that is easy to recognize and works on many of the same principles as ours—with only small changes.

My goal each time I design a new game is to create a world that requires only the smallest suspension of disbelief for the player to believe it is real. The player has only the tools we provide to solve the problems we put in front of them, so it's important that anything in a game that looks like it ought to do something must do something: If there is a chair, you ought to be able to sit in it. If there are lights, you should be able to turn them off and on. If there is a refrigerator, there ought to be food in it that you can eat. If there is a telephone on a desk, a player has to be able to pick it up and it has to function—even if it just rings and gives a busy tone. If there is a harpsichord, it has to produce music. If there was a door or window, it had to be able to be opened or locked. To me that is an essential aspect of gaming, one that sets my games apart and has led to me being publicly and loudly critical of many game designers.

This belief goes all the way back to Dungeons & Dragons.

Before D&D the only games I knew were board games like Monopoly in which players follow a strict set of rules, and the only way to win is to follow the instructions. If you draw a Go Directly to Jail card, you have to go directly to jail. You can't plead for a lesser penalty, you can't hire a great lawyer, you must go directly to jail.

While initially D&D did have a rulebook, the rules mostly were story-telling guidelines. And most players ignored them. In fact, if you look at the origin of D&D, the rules say that they are written for people to play with as if they were miniatures, like toy soldiers. These were fantasy toy soldiers and the rules of engagement were pretty strict; for example, when your toy soldier is one inch away from the other toy soldier, you roll the dice to determine the damage. The original rules weren't written as a platform to encourage interactive storytelling, but the early adopters ran with it and took the game in an entirely new direction. By the time I first played D&D, the storyteller would completely create the environment: there's a table over there, a chair over there, and the monster is coming through that door. The miniature rules had been abandoned: You have a knife, a fork, and a spoon, and there is a glass hanging on the wall. What do you do?

As long as the answer was funny or clever, or would have worked, or could have worked, or at least made progress, the gamemaster would find it acceptable. But if the thing you did was stand around and do nothing, or do something that in the mind of the storyteller made things worse, then guess what, it didn't help. That's how the game was played without rules. It was a negotiated narrative.

This was the essential foundation for my games. Many competing games could reproduce what a world looked like—but not how it functioned. Few items were interactive. In most games if there was a sink, you couldn't turn on the water. The lights were permanently on. Peripheral characters in those games existed for the sole purpose of telling you something or trading with you. Players had few options, and as though they were playing Monopoly, they had to follow the rules.

When I design a game, I want people to have that same ability to creatively solve problems that they have in D&D. If we confront the player with a problem, they should be able to solve it in a variety of different ways, rather than being forced to guess the one solution the developers have in mind. Often in D&D players come up with solutions the gamemaster hasn't considered, and I always want to give players of my games that same capability.

This was the metaphor I used when writing a game: Imagine that we are locked in an office and the building is on fire. What steps can we take to save our lives? Can I smash the lock? If the door is wood, is there something in the room I could use to smash through it? If I break a window, can I get down to the ground safely? If the door hinges are on the inside of the room, could I knock out the hinge pins and pull the door off backward? If I couldn't go through the door, could I go through the ceiling? If it's an acoustic ceiling, could I push through some of the tiles and go over the wall? If the walls are made of Sheetrock, can I break through them? The point is that everything in the room is potentially a tool to help me—but only if the environment is completely interactive. In most games the player's imagination is limited by the creators: the only way to save yourself is to do what the designers want you to do.

Rapidly improving technology and gains in computer memory made it possible for me to create an increasingly interactive environment with each new *Ultima.* I eventually came to believe that instead of conceiving of a story and then creating a world in which it could be told and make sense, I should create the world first—then let the player have access to all of its capabilities as they negotiate their own story. The door did open; it was up to the player to decide whether or not to open it.

Certainly one of the most difficult challenges we faced in each succeeding game was to produce a worthwhile villain, one with enough cunning and guile to make the player feel they were facing a real and very capable enemy, an antagonist worthy of the player's time and concentration. In too many games the bad guy is obvious and pretty easily defeated. It may be a horrifying monster or a bloodthirsty outlaw, but unintelligent. The only reason for its existence is to be defeated. In my earliest games, when the graphics were very primitive, an icon would represent the monster. It wasn't much of a monster, it just waved its arms and yelled, "Arrggghhhh," and the player hit it repeatedly until it died.

I found that when it comes to monsters, bigger isn't always better. When you're creating a pantheon of opponents in a virtual world, it's easy to make small monsters, like rats, that then can transition into bigger opponents, like wolves, which become bigger trolls, and finally a dragon. However, if all the monsters do is become bigger and tougher, then the player's behavior isn't really tested. They just trade blows with increasingly bigger opponents until they win or lose. I wanted to challenge players in a new way.

To accomplish that I had to create monsters whose characteristics—not their size—would demand a more challenging response from the player. For example, I enjoy putting creatures like slimes in the game. If you attack them with a sword, it causes them to split in half, with each half as strong as the original. So if you attack them in the traditional way, your problem only becomes worse. But if you attack them with a torch that dries out their moist bodies, you can defeat them. We had a lot of fun playing with the possibilities: maybe each time they split, instead of being as strong as the original, they get a little weaker; maybe each half is only three-quarters as strong, which means if you keep splitting them you can eventually get to a point where they are so small they can be easily dispatched.

In the first three *Ultima*s, the antagonist was a somewhat traditional bad guy, or woman, although even then we were at least trying to make our monsters more involved in the storyline. Essentially, though, these were standard role-playing games. In *Ultimas IV, V,* and *VI,* there was no bad guy, rather, they focused on the player's personal pursuit of enlightenment. Eventually though, I sought a way to combine the two models, integrating both virtue and more complex villains who were thoroughly incorporated into the story.

So when we began designing *Ultima VII,* I needed to create an enemy that was worth the difficult pursuit. I needed to create the Best Possible Bad Guy!

This was also the first time I looked beyond the end of the specific game I was working on. I thought it would be interesting to create a dark force, an ultimate opponent that would appear in the next three games, and would tie them together as a trilogy.

This ultimate force of evil would be called . . . The Guardian! Players would meet and defeat his most visible minions in *Ultima VII,* but the ultimate showdown with him would not take place for two more games. In fact, in *Ultima VIII* The Guardian would gain the upper hand and challenge the player to survive in a world ripe with absolute evil.

I spent a great deal of time creating The Guardian. It couldn't just be some kind of monster with incredible powers, it had to be an evil entity that might truly exist.

I found the answer in a newspaper insert. At the time the Church of Scientology was being investigated by various government agencies. To combat that bad publicity, Scientology circulated a sixteen-page full-color insert tucked inside *USA Today.* The insert included a photograph of Scientology founder L. Ron Hubbard on the front page with a caption that described him as "a Renaissance man," and each subsequent page covered his achievements: at thirteen America's youngest Eagle Scout, aviation pioneer, explorer, mariner, and World War II captain. Each of these was illustrated with some type of inspiring photograph: Hubbard the master mariner at the helm of a ship; Hubbard the philosopher sitting at his desk with a quill pen; Hubbard the humanitarian feeding children. There were sixteen pages of this. And one of the last pages boasted, which I thought was completely ironic, "L. Ron Hubbard, one of the greatest writers of pulp fiction." Correct, I thought, lots and lots of fiction. The final page was a call to Dianetics.

The insert was a valiant attempt by a PR agency. While I didn't take any of it seriously, I knew that a lot of people did. It dawned on me that this was the type of enemy I needed to create. If you were going to create a charismatic leader, I realized, these were the qualities you would want him to have. I thought about the charismatic leaders in history; with the exception of recent madmen, none of us knows what they were actually like. No one knows what the real Buddha enjoyed eating, for example, or if Moses had a sense of humor. Over the course of time, their cult status prevented us from seeing them as actual human beings. But watching it happen to a contemporary figure like L. Ron Hubbard was fascinating. This was the pattern for creating a god, and out of it was born Batlin, the chief villain of *Ultima VII*.

Batlin was never intended to be the ultimate evil, but rather the puppet of the greatest evil in this world, The Guardian. The Guardian is a malevolent force that transcends the worldly plane. What made both Batlin and The Guardian especially evil was their being fully aware of the harm they were causing. Purposely hurting other people was the way they got ahead. Batlin is not Hubbard, nor was he intended to be Hubbard in any way. I was more interested in how Hubbard was promoted as a great leader than in the specific details of his character. In fact, I felt I had to invent a whole new philosophical code for this game, something that this leader could embrace and expound upon that went further than truth, love, and courage. It had to be simple, yet meaningful enough to attract a following who would believe it was the truth—and it had to be a set of values that I could later twist into something really, really dark. Eventually I settled on unity, trust, and worthiness.

Batlin was the leader of an organization called The Fellowship that traveled through the land recruiting the good people of Britannia to join them around these three ideals. The way the leaders of The Fellowship would explain them publicly was that we, as a society, needed to band together because we could accomplish so much more as a tribe than as individuals. To join together on that path, we had to trust each other, and each of us should strive to be worthy of membership in this tribe. Through these ideals, we would individually perform better, find the freedom to operate independently, and ultimately the tribe would advance. It seemed like a positive and reasonable message. Just like a lot of the tenets of Scientology.

But then I began twisting it into something considerably darker. Unity, for example, eventually came to mean you were either with us or against us. Anyone who wasn't part of The Fellowship became the enemy. Trust mutated into a belief in everything Batlin said; no one could question his authority. And adherents had to keep proving their worthiness to the tribe, at the expense of all else.

And that is how the positive-sounding Fellowship could transform into something sinister.

In *The Book of Fellowship,* Batlin also very subtly discredits the original eight virtues. While they are certainly a worthwhile goal, he writes, they are too hard for most people to follow. Instead, The Fellowship offers a simpler path to reach enlightenment. In fact, he was creating a "new" truth.

It was all designed for the player to believe, at least initially, that Batlin was trying to spread positive values. I hoped players would want to enlist in The Fellowship, thinking that these were the new virtues of the game. It was not until they got further into the game, after they had ceded control and given public support to this movement, that they saw the brutality and the darker interpretations of these same virtues.

The true purpose of The Fellowship was to create the gateway that allowed an evil entity called The Guardian to get inside Britannia and eventually gain control. All of that background helped make Batlin a believable monster, allowing the player to fight for the existence of Britannia.

Trust me on that.

9 CREATE

The Haunted Beginnings of Themed Interactive Events

Growing up in Houston, just outside the front gates of NASA, I thought most everybody's father went into space. Mine, Owen Garriott, was a scientist-astronaut who in 1973 spent sixty consecutive days in space aboard Skylab, which at the time set the record for longevity in space. Almost everyone in our neighborhood worked for NASA in some capacity; my next-door neighbors on both sides were also astronauts. When I walked the dog, I went past the homes of astronauts, scientists, mathematicians, engineers, and other people who spent their lives creating the future. It was a place rooted firmly in science and the exploration of the reality in which we live.

As a young child, I didn't pay a lot of attention to historical fantasy, mythology, or the supernatural. My real life was populated by courageous explorers, risking their lives practically every day.

At least I didn't pay attention to it until the Halloween that changed my life. I was eight years old and determined to collect all the candy in the world. Many of the houses in our neighborhood had carved pumpkins out in front or cardboard skeletons suspended in the windows and some plastic cobwebs covering part of the front porch, but one house stood out. It was on the other side of the street and none of us knew the person who lived there, but the house had been elaborately decorated, far beyond what anyone else in our neighborhood had ever done. We crossed the street and rang the bell.

A witch answered the door. From her curled-toe shoes to the tip of her black pointed hat, she was the witch of our nightmares. She wasn't wearing the usual store-bought costume, but rather an elaborately conceived black dress. Her makeup and prosthetics went well beyond the other homes we visited. She had a long, thin, cursive nose, sharpened fingernails that looked like claws, and her skin was made up to look a hundred years old. "Trick or treat?" we said, but probably not too convincingly.

"Come in, my little chickees," she cackled. This was the first time that evening anyone had invited us into their home; it was also the first time I had been in what was clearly a real witch's den. The darkened interior was lit mostly by candlelight, several large spiders were sitting in webs hanging from the ceiling, and haunting music played in the background. It was as perfectly detailed as her outfit. And placed right in the center of the entrance was a large, boiling, gurgling cauldron. She held a violet wand in her hand that was about the size of a man's electric razor with a long glass tip attached to it; it looked incredibly scary and dangerous. When she touched you with the tip, it crackled and emitted sparks and created a slight tingling feeling.

Before we were allowed to take a handful of candy she had to examine us with her wand to determine if we were worthy. At that age, in that house, it required sincere bravery to succeed at this mission. I later learned the wand was a harmless Tesla coil, but the effect was tremendous. The whole scene was truly in-

TOP: "The Life of the Cicada Killer," my kindergarten science fair project at age five. My mother encouraged me to follow a bug in our backyard, which turned out to be a fascinating creature that paralyzed cicadas, then laid its eggs inside of them. A lifelong curiosity, and love of exploration, was born.

MIDDLE: My younger sister, Linda, and I were best friends growing up, and I often shared my magic with her.

BOTTOM: The Garriott sibling clan. One of the rare occasions that saw us in formal attire. Must have been a wedding!

The leather jacket days: looking "tough" in front of a poster of *Ultima III: Exodus.* When I moved to the northeast in 1984, I entered my black phase. This was also my first braid, which I grew behind my ear. I got tired of people asking me if I was wearing a hearing aid, so I cut it off and grew a new one in the back.

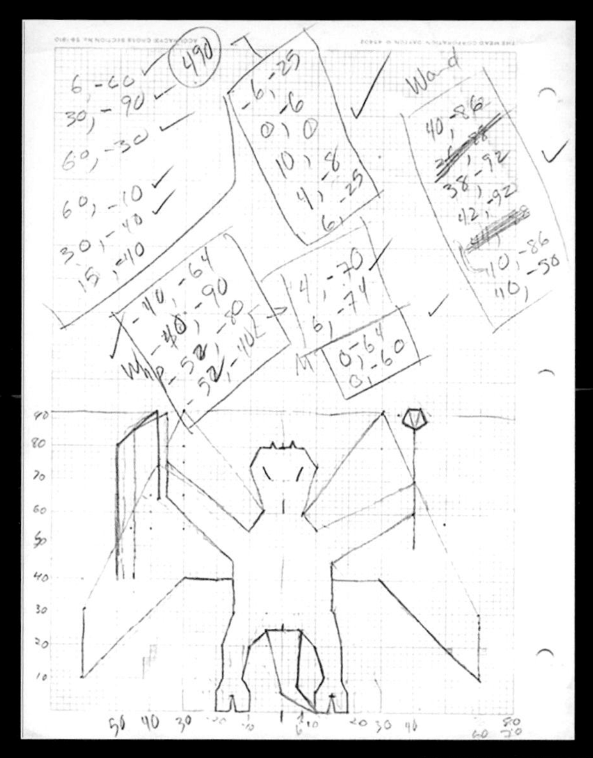

Akalabeth was the first 3-D roleplaying game, published in 1980. Here is my early rendering of the monster, a balrog (borrowed from Tolkien). This page outlined some of the first ever 3-D art, drawn out on graph paper and hand-calculated into perspective.

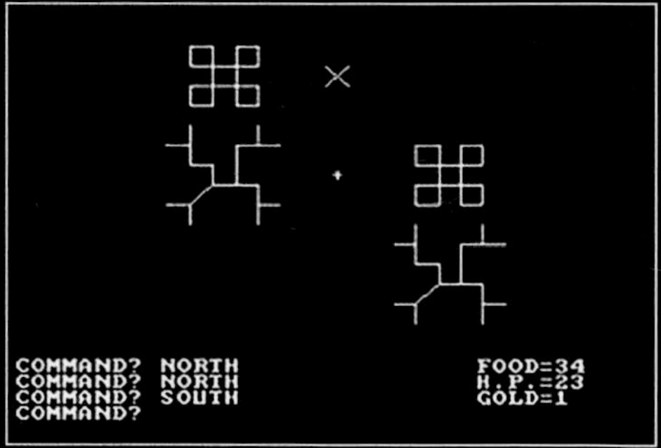

A bird's-eye view of the outdoors in *Akalabeth*, showing a dungeon, two towns, and two mountain ranges. Graphics have come a long way since!

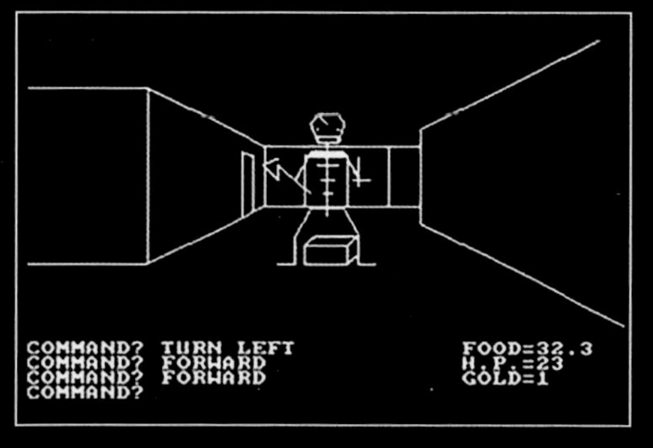

A dungeon in *Akalabeth*.

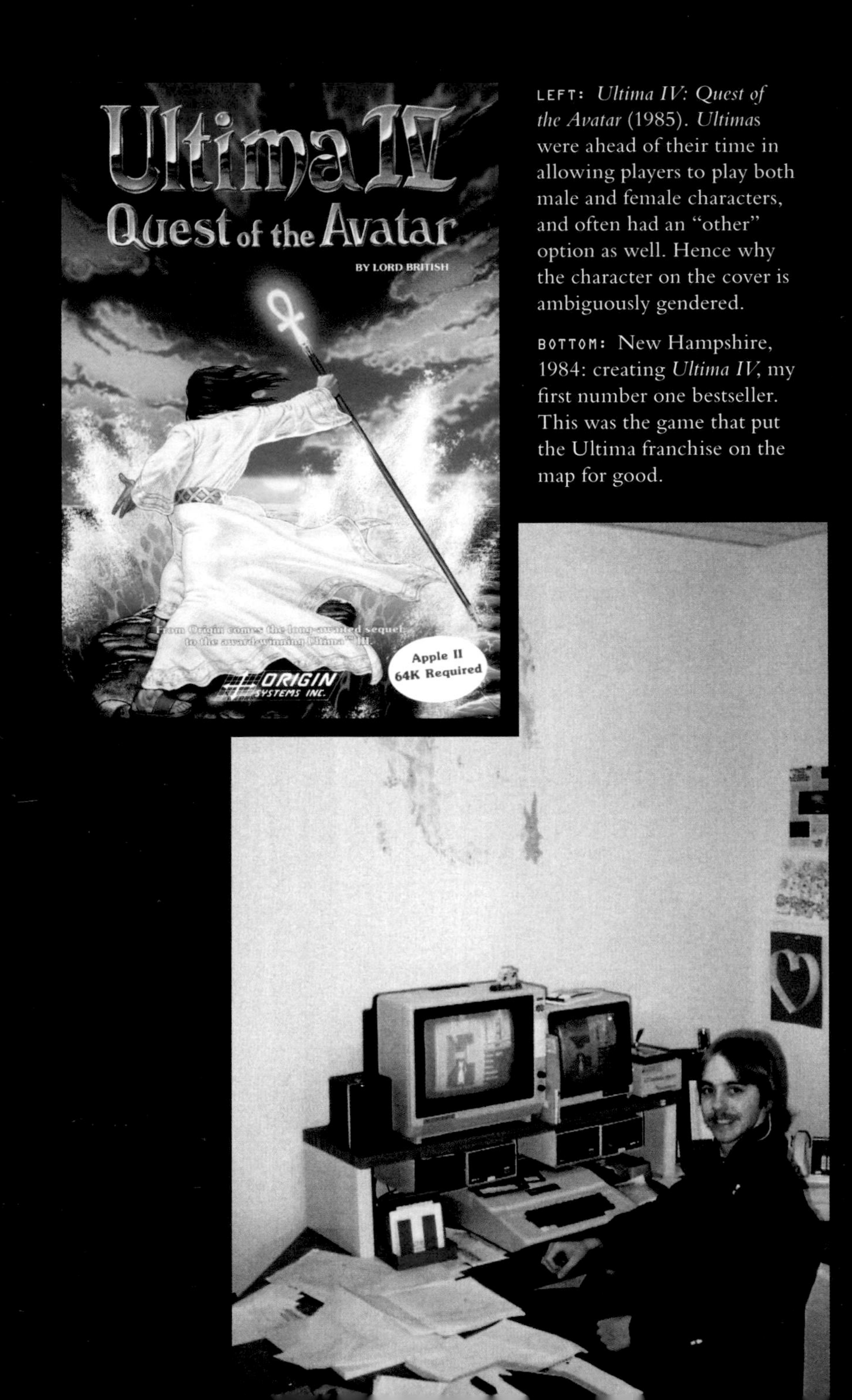

LEFT: *Ultima IV: Quest of the Avatar* (1985). *Ultima*s were ahead of their time in allowing players to play both male and female characters, and often had an "other" option as well. Hence why the character on the cover is ambiguously gendered.

BOTTOM: New Hampshire, 1984: creating *Ultima IV*, my first number one bestseller. This was the game that put the Ultima franchise on the map for good.

TOP: Lord British and Blackthorn addressing the masses on our grand tour before the beta servers for *Ultima Online* were wiped clean. What a crowd!

MIDDLE: One of the most infamous moments in video game history: the assassination of Lord British. Rainz, one day we will meet again!

BOTTOM: Crime never paid in *Ultima Online*. Here, a crowd gathers to watch a public execution.

TOP: Britannia Manor, my old residence in Austin, the site of my legendary haunted houses and once home to many of my collections.

MIDDLE: I have a particular fascination with orreries, as they combine my love of astronomy and mechanical models.

BOTTOM: It's no surprise I love medieval arms and armor. This piece, a particular favorite, was worn in the last jousting tournament of the Victorian age.

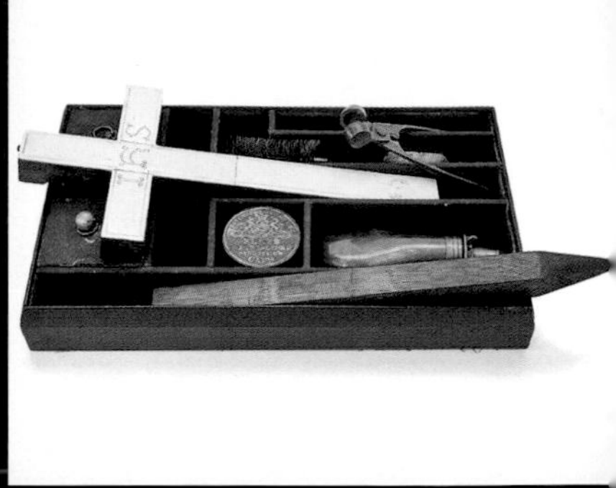

TOP LEFT AND RIGHT: One of my more unusual collections: vampire hunting kits. Here are two shots of my favorite, which includes the traditional cross and stake, and a variety of oils and elixirs. These are more than a century old, from a vampire craze that far predates *Twilight*.

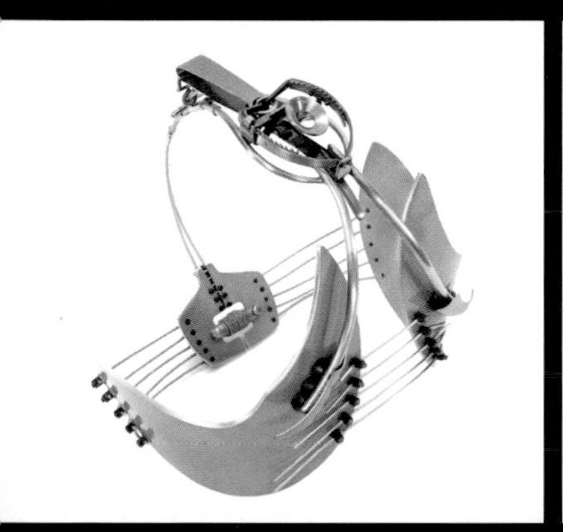

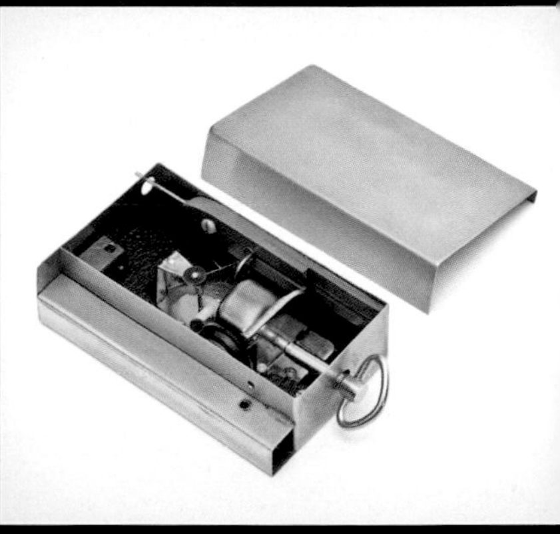

MIDDLE LEFT: Item 1

MIDDLE RIGHT: Item 2

BOTTOM LEFT: Item 3

MIDDLE LEFT, MIDDLE RIGHT, BOTTOM LEFT: Here, readers, are three mysterious items from my collection. Think you know what they do? Tweet @RichardGarriott with your guesses and the item number.

TOP LEFT: At the 1998 Britannia Manor haunted house. The flames behind me are an oil well fire simulator.

TOP RIGHT: If there is any doubt where I got my penchant for the macabre, this drooling gargoyle fountain was made by my mother.

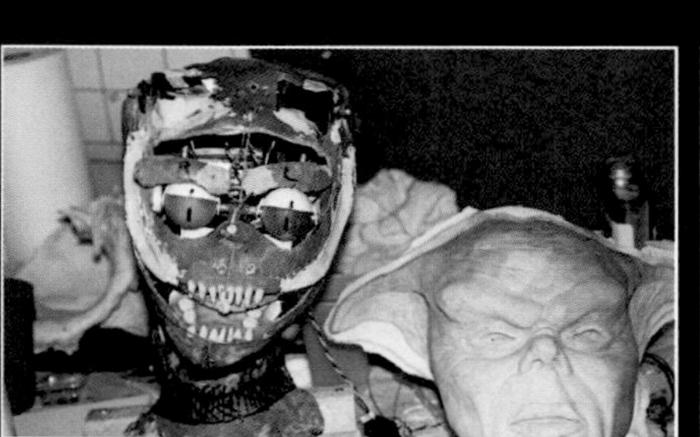

MIDDLE LEFT: My volunteer team orchestrated sophisticated prosthetic animatronics. Many of them went on to careers in Hollywood.

MIDDLE RIGHT: During the haunted house, visitors had to cross the River Styx at midnight while paying the ferryman a dear price.

BOTTOM LEFT: The demonic finale of the Britannia Manor haunted house.

TITANIC SINKS!

TRAGIC LOSS OF LIFE

TITANIC SINKS FOUR HOURS AFTER HITTING ICEBERG; 866 RESCUED BY CARPATHIA, PROBABLY 1250 PERISH

Biggest Liner Plunges to Bottom

RESCUERS ARRIVE TOO LATE

Manager of Line Insisted Titanic was Unsinkable even after She had Gone Down.

FROM THE WHITE STAR LINE

With profound sadness we must announce the recent death of Mr. Scott Jones, one of the passengers on the ill fated Titanic. His body was recovered and buried at sea by the Cable Steamer 'Mackay-Bennett'. It is believed that he drowned.

Those wishing to pay their respects may do so at a memorial service being held in honor of all those unfortunates who persihed in this tragic event. The memorial service will be held Sunday April 21st, 1912 at Saint Paul's Church in Halifax, N. S.

Richard Allen Garriott
MANAGER, WHITE STAR LINE

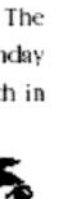

News clipping for a lost soul.

LEFT: The "news clipping" from my *Titanic*-themed dinner party. Sadly, this particular passenger did not survive the wreck.

BELOW: I too, of course, went down with the ship, along with Austin mayor Kirk Watson and most all of the other men.

The Captain

BELOW: A different sort of aquatic amusement. Our winning entry for the Austin River Raft Race art contest: a Nautilus submarine, a giant squid, and a steamship that would break in half and sink, all as we floated down the river.

Staring down the bow of the *Titanic*.

LEFT: Visiting the wreckage of the *Titanic*. This is an upper deck cabin right at the fracture point; you can see on the left edge where the cabin ends at the point where the ship was ripped in half.

RIGHT: On the deck of the *Titanic*. Significant debris has fallen onto the deck, including some wall panels.

LEFT: The seafloor is especially littered with debris that fell out of the *Titanic* during its separation and descent: toilets, benches, plates, cups, personal effects. All make the human tragedy feel a lot more real.

ABOVE: My first expedition to Africa—really, my first expedition anywhere. I'm reading *Hubert's Hair-Raising Adventure* by Bill Peet, a story that is deeply meaningful to my whole family. I have memorized it and take it on all my adventures!

BELOW: The most inhospitable place on earth: trekking through Antarctica, land of meteorites.

LEFT: Ringside for Jesús Chávez vs. Wilfredo Negron in Austin, 1997.

LEFT: The view from the corner: watching Chávez box against Sirimongkol Singwancha for the WBC World Super Featherweight Championship, 2003.

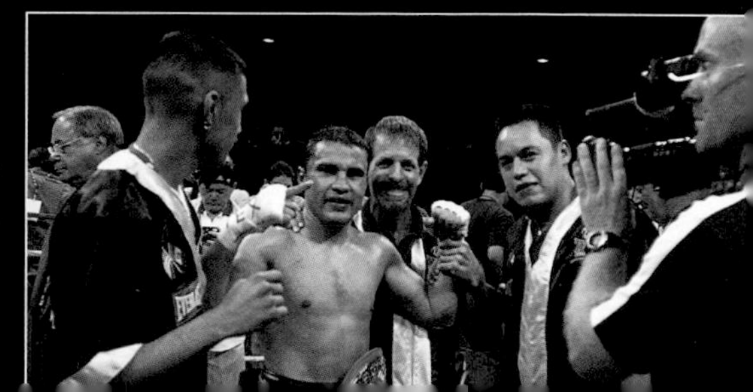

RIGHT: Chávez with title belt after his victory, explaining to reporters the Garriott good luck charm!

LEFT: Posing in front of a photo of my father, Owen Garriott, an engineer and astronaut who spent sixty days in space on the Skylab II in 1973.

BELOW: My father doing a spacewalk during the Skylab II mission. *(NASA)*

RIGHT: Sharing stories of my flight with Dad, mere moments after returning to the earth.

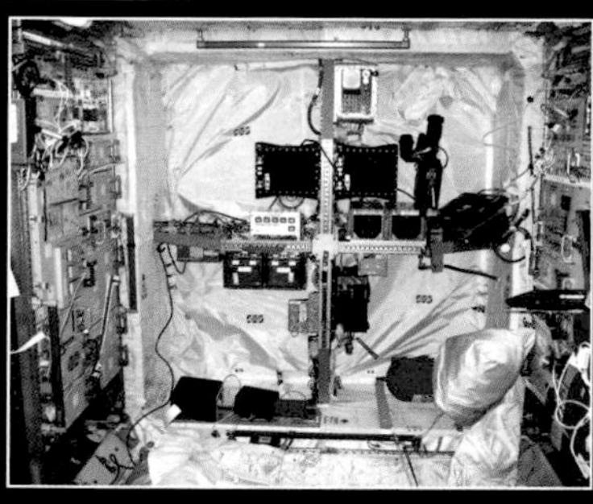

TOP: The official crew photo prior to our launch. Mike Fincke is on the right and Yury Lonchakov is in the middle. While NASA refers to me as a "space flight participant," England is happy to call me an astronaut (I'm one of only half a dozen British-born people to go into space). Hence why I'm rocking both the Stars and Stripes and the Union Jack.

MIDDLE LEFT: The digs. Not quite as luxurious as Britannia Manor, but comfy. Hidden somewhere in here is the secret resting place of James Doohan.

MIDDLE RIGHT: Our crew press conference from the International Space Station.

RIGHT: Formula 4001, Stephen Colbert's DNA that I took to the ISS with me. If we one day find intelligent life forms out there, we'll know where they came from.

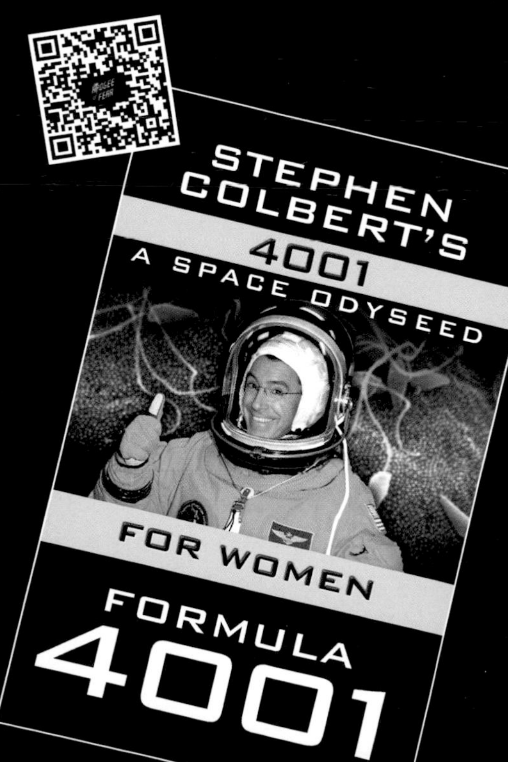

RIGHT: With my wife, Laetitia, on our wedding day. We floated magically up into the sky for our kiss.

ABOVE: Laetitia and I with our children, Kinga and Ronin. "Kinga" is derived from a word meaning "loved by everyone" in Bhutan, the land of Gross National Happiness that we explored in 2010. "Ronin" is drawn from a Japanese word translated as "samurai who has no master" and we chose this name for him because, to us, it projects the idea of a powerful and independent spirit who—because he doesn't have a master—doesn't fight for a master, but rather for the greater communal good. Their middle names are symbols. Kinga's is 水龍, which means "water dragon," from the year of the Chinese zodiac in which she was born, and Ronin's is φ, the "golden ratio," an irrational number commonly found in the natural world.

LEFT: Come explore further . . .

timidating, but not in any inappropriate way. It was an entirely visceral experience. There was no real threat, no torture, no one touched us or threatened us—but it required courage to work through this audiovisual sensational experience to get our candy and get out of there.

Even though I was a bit afraid, internally I was also smiling widely because I immediately recognized the power of everything that was going on around me. I knew we were in a real world and there was no actual danger, but I was wonderfully intimidated. She had put together something so far beyond what I could possibly have conceived of that I was superexcited and superaware of everything going on around me. And I thought, If somebody else can make me feel this way, I want that power. I wanted to share this wonderfully joyful sensation I was feeling with everyone around me. If there is a single pivotal event in my life, this would be it. It's fair to say that almost everything I've done since then has reflected that desire.

The next Halloween I told my mother I needed to decorate our house. The result was acceptably scary, but didn't come close to accomplishing what I had in mind. What had impressed me most about the witch was not necessarily the house decorations, but the interactivity of it. That became my goal. I found several books that offered suggestions. For my first haunted house, I created several tactile experiences that people had to go through before earning their treat. For example, I put grapes in a bowl of ice water and called them eyeballs and visitors had to reach in and touch them. It was a start.

The following year I explained to my mother that in addition to the front porch I needed to use the entryway. This was my project; my older siblings were ahead of me in school and wanted no part in trick or treating and my parents were just pleased I was keeping out of trouble—as long as I cleaned up afterward. After

I had taken over the front porch, entryway, and living room, my mom pushed me out into the yard; that was when my visions really began to take shape.

Our backyard had two play forts, my father's aluminum tool-shed, and a compost heap, and I used all of it. The forts became vignettes of frightening experiences, encounters with a witch or a soothsayer, and my father's toolshed was the blood-covered hide-out of a maniacal murderer. We set up a ritualistic scene with candles around the compost heap, and as visitors moved closer to see what it was, suddenly zombies, screaming loudly, popped up from beneath the pile of leaves.

Creating that experience was so much fun I began doing it year-round. I built a haunted house for Easter, and I may have been the only person in America who celebrated the Fourth of July with a haunted house. Gradually these haunted houses became more and more elaborate. I always wanted to push the bounds of innovation.

Commercial haunted houses are decorated like movie sets and have people dressed as monsters or bloodied victims jumping out of dark recesses and screaming. The goal is to be acceptably scary and run as many people through as quickly and profitably as possible. But I was more focused on creating a memorable interactive experience. I never charged admission to my haunted houses, and instead of volume, I focused on quality.

As far as I know there had never been anything like my haunted houses in Houston. I changed the rules. Rather than it being a mostly passive series of scares, people going through my haunted house were involved in a little quest. They had to climb over obstacles, swing across chasms, unlock "multi-user" latches, and even negotiate with characters. They had to find some sort of artifact that was their magical key to getting past another barrier. Failure, as well as success, was possible. It was as personal an experience as I could produce; instead of the usual continuous line of people following a leader, people went through my haunted houses in small groups, and until one group was far enough through that letting in another group wouldn't affect their experience, no one else would be allowed inside. Eventually, though, the events got too big. I ran out of room and my mother's good graces, so I stopped.

Several years later my brother and I formed our first video game company and settled in New Hampshire. I owned my own home, so I could do whatever I wanted with it, and I began creating bigger and better haunted houses. While initially I was trying to entertain young trick or treaters, I discovered that their parents enjoyed it as much as they did, and in fact eventually adults outnumbered kids. It was even more difficult satisfying such a wide age range.

After three years in New Hampshire, we moved to Austin, Texas. The slogan proudly used by people who live there is "Keep Austin Weird," so it was the perfect place for us. Many people have used their financial success to build their dream house; I used mine to build my dream haunted house. It was in Austin that my haunted houses became world famous.

I've approached every creative project throughout my career with this same goal: to create an experience beyond anyone's expectations, and involve every participant as completely as possible in the reality of my fantasies. My desire has always been to introduce something so profoundly different that it forces people to pause and be completely in the intended emotion of the moment, something so awesome that everyone will have to give it their complete attention, meaning that for a very brief time all other goals and thoughts and problems just disappear.

People go into a haunted house confident that nothing actually dangerous is going to happen to them. They expect to be surprised and startled, but they figure it's comparable to watching a horror film—they do not expect to experience actual fear. The knowledge that the experience isn't actually dangerous prevents them from participating emotionally, even if a bloodied character jumps out of the dark screaming in their face. My goal was to get beyond that and make a clearly artificial experience as real as possible.

It took me and my team as long as six months to design and construct each set, and costs grew into the hundreds of thousands of dollars to stage each one, so I only put on the haunted house every other year. To make sure every participant had the best possible experience, we limited it to 250 visitors a night, on a first-come, first-serve basis, and opened it for only four nights; scaring a thousand people is a tremendous amount of work. As my haunted house became widely known, people would camp out in front for two weeks before we opened to make sure they got in. Visitors came from all over the world. We often had celebrity guests like Mark Hamill and John Rhys-Davies. Some people enjoyed it so much they volunteered to work at the next haunted house, which was very important, because at times it required hundreds of people to operate all our illusions.

Creating haunted houses has recently become a profitable business, but at that time mine was the most elaborate and terrifying haunted house in existence. Because I lived outside the city, I knew we wouldn't have neighborhood trick or treaters in plastic costumes, and in fact almost all of my visitors were adults. Adults don't want to act scared, they want to seem cool and in control of their environment. Of course they will often instinctively protect themselves; if a monster jumps out in front of them they might respond by punching the monster. Adults are control freaks, especially men. Men want to show that they are in charge.

Oh, believe me, they were not in charge. I discovered that if I took control away from them, even the most stalwart of adults would feel significantly ill at ease. We found that we couldn't do that with actors—but we could with unique mechanical devices.

We began by constructing an entire false front to my home that made it look like a collapsed ruin. Just to get inside, visitors had to climb over and crawl under fallen timbers. We made it easy enough so that no one would be unable to enter, but difficult enough that people would understand this was not going to be a passive walk-through. They were going to have to work for their frights.

We created an environment that might accurately be described as a prototype of hell. In my sand volleyball court we invited the local volunteer fire department to create a pyrotechnic display. We weren't interested in spectacular fireworks or burning oil barrels that people could admire from a distance; we intended to put them right into the middle of it. We had two fire effects in the sand pit. One was an oil well simulator that shot doughnuts of flames straight up into the air. But to keep those pipes clear when no fuel was being pushed through them we had to have a steady flow of water. As a result, when flame balls shot into the air, it also created a mist of moisture, which smelled like gasoline and successfully created the illusion as people walked through that it was raining fire. Those droplets of water and gas fell on the sand; the water would be absorbed but the fuel stayed on the surface and eventually caught fire, so there was a constant silken veil of flames dancing all over the volleyball court. It wasn't physically dangerous, but the sensory impression of raining flames was fantastic.

This was the intersection of science and the senses. The emotional experience was completely detached from the physical experience. The senses were sending very different messages to the brain than the body was actually experiencing. No matter how strongly people believed that they were safe, no matter how well they knew intellectually that no sane person would actually put them inside a boiling rain of fire, their senses were sending very different messages. Their sense of security was destroyed. Imagine standing in pools of flame literally dancing around your feet, while only a few feet away great fireballs are rising from the highest point in Austin, sixty or seventy feet into the air, so high that they are visible for miles, from any place in the entire city. From the outside, it appeared miraculous that we didn't kill anybody, but there was little real danger.

The second effect consisted of the propane jets used in hot-air balloons that create very precise columns of flame; properly positioned you can use them to create a horizontal wall of flames. We used those propane jets to make walls of flame around the volleyball court. Visitors had to navigate their way through these flames to a raised dais built of human skulls and body parts where Beelzebub was waiting, dressed in the regalia of great fantasy, his voice modulated through powerfully loud speakers. The entire fog-shrouded scene, which included walls of flame, fireballs bursting into the air, and flames on the ground created a perfect atmosphere.

After surviving the rain of fire, visitors made it to the safety of my house. Safety?

Hardly, and that is where my true diabolical nature took control of them.

Inside the house, visitors had to crawl through an indoor pool that had been decorated to resemble a sewer, and eventually make their way to a walkway that floated on the surface of the outdoor pool. And just as I had done years earlier, as people walked across the fog-covered pool, monsters reached out from the water and tried to grab them by their ankles. And that was just the beginning.

Eventually everyone was funneled into a totally black maze. They entered this maze in groups of four. We noticed that they often held hands as they eased into it.

Did they really believe holding hands would save them . . . from me?

At the end of one long corridor, visitors were led into what initially appeared to be a dead end, but was actually a narrow hallway about eight feet long. In fact, it was so narrow that people had to go through it sideways; by design it was impossible for two people to walk through it side by side. When all four people in a group were in that section, boom! Suddenly, the walls started moving inward, coming closer and closer until everyone was trapped between them. No one could escape. And then the walls began squeezing them. What they did not know was that those realistically appearing walls were actually two-foot-thick layers of absorbent foam. Even if someone reached out awkwardly to try to stop the walls from crushing them, they wouldn't be hurt.

Once we were certain the whole group was safely encased within the foam walls, pneumatic cylinders behind those walls literally lifted the room off the ground, flipped it over, and slammed it back into the ground. It was accompanied by very loud and grinding sounds and it took place in total blackness.

It was completely disorienting; people really feared they could be injured, but there was nothing anyone could do to stop it. They had lost control. Eventually the pneumatic cylinders would separate the walls, allowing everyone to scamper out through the gap between them. Oh, how they scampered.

In another segment our diabolical goal was to identify the most easily scared person in each group and torture him or her personally. We built a twisting tube that forced everyone to get on their hands and knees and crawl through single file, meaning once again it was impossible to hold hands. There were holes in the tube so we could watch their progress, and we had the ability to drop gates in front and behind any individual in the tube. Usually we'd wait until the brave leader had passed through it, then we would drop the gates around their scared friend. While the others were in different segments so they couldn't see this happening, we would slide out that section of tube while instantly replacing it with an empty section then raise the gates. Three of the four people who entered the tube would come out into what looked very much like Dr. Frankenstein's laboratory. But one person with whom they had entered the tube only minutes before—had disappeared!

On the far side of this madman's laboratory was a giant Tesla coil; as the three now disoriented people stood there, a sheet would be dropped to reveal the fate of their fourth friend—who was trapped inside a cage! Although none of our guests knew it, this was actually a disguised Faraday cage, essentially a chicken wire enclosure. It allowed us to create an amazing illusion: Electricity would flow around the perimeter into rods in the ground, but none of it would go through the wire into the cage itself. It works for the same scientific reason that people remain perfectly safe when the car or airplane in which they are riding is hit by lightning. As Michael Faraday discovered in 1836, the energy is carried only on the outer perimeter of the object because of the repulsive effort of electrons inside. The person inside the cage was completely safe, although from the outside it certainly didn't look that way.

That's when the real fun began. Our mad scientist would scream insane threats, then turn the large dials of high-voltage machinery, including the multimillion-volt Tesla coil. Suddenly bright, hot, and seemingly deadly sparks would appear to ignite the tiny cage in which the most timid member of the group lay trapped—and often, screaming. It was a wonderfully loud, amazingly bright illusion; you could even smell the unmistakable odor of ozone. It was not at all unreasonable for the person inside that cage to think they were going to die. A thought, by the way, that we greatly encouraged.

Even above the crash of our lightning and the grinding of our machinery, we could hear their screams. It was the scariest creation we had ever had in a haunted house. We knew that for certain because this was the place where the most people wet themselves, which was our measure of success.

During the years we operated the haunted house, we had only one potentially serious problem. One year we included what we called the crazy cart, in which we put people in a black box and drove them down a steep hill in a rickety vehicle. One night, while the car was careening sharply back and forth, a door broke off and one of the passengers fell out of this crazy cart and onto my driveway. I think we all held our breath as he lay there, as he could have been badly injured; instead he popped right up and took off running after this cart, screaming for them to wait for him.

While I can claim credit for many of the ideas, it was my friend and partner in these adventures, Greg Dykes (also the model for the character Dupree in my *Ultima* games), who was foremost in turning our concepts into mechanical reality. We also used really good special-effects designers and makeup artists for our actors. At the end of every evening, after we had shut down the power, the entire 220-person team gathered in my front yard, opened some beers, and told the stories from the night. How many people were so scared they couldn't make it to the end? What got broken? And, of course, how many people wet themselves?

While I could easily imagine my haunted house getting more elaborate, it had reached the limits of practicality. From having a costumed friend pop up from beneath a pile of leaves I had taken it to a scale that required most of a year and a lot of money to build and tear down. It didn't make sense to make it bigger or bolder without turning it into a business, and doing that didn't excite me.

So while that marked the end of my haunted houses, it also marked the beginning of my creating interactive events. It was Pablo Picasso who once said, "Creativity is the tool your mind uses to scratch an itch in your imagination." I believe that quote is entirely accurate, otherwise I wouldn't have made it up.

Creating Reality

A lot of people have tried to define creativity without much success. Maybe the single most accurate definition of creativity I've ever read is the one that appears on page 125 of this book. Generally though, creativity seems to fit the United States Supreme Court's definition of pornography: there is no universal definition, but you know it when you see it.

Creativity isn't something you can buy, it isn't a possession one person can give to another. Science doesn't know precisely where it comes from or how it develops. It just is. In my life creativity is the added and usually unexpected element that transforms the ordinary into the memorable, a spice that makes the world more interesting, and most of the time, a lot more fun. Truthfully, how many people have to hire stuntmen and animal wranglers for their own wedding?

My need to create has also driven me to put new twists on everyday events, like the dinner party. I can't simply host a party, it has to be a memorable, interactive event. People who know me know that when I invite them to something, they are not simply going to observe; somehow, they will have to be active participants. They know to expect the unexpected; in fact, just about the only thing I could do at an event I hosted that would be completely unexpected would be to do nothing unexpected. People would be shocked: What's the matter with Richard? Everything's normal. Which certainly was not the case the night I nearly drowned the mayor of Austin.

After returning from my *Titanic* trip, I held a themed dinner party at my house in Austin: A voyage on the *Titanic*. In addition to my friends, I invited a cross section of the city's leading citizens, including the mayor and his wife, members of the city council, and local business leaders. Everyone invited to this party had to decide in advance if they would be traveling in first class or steerage. Those who chose first class had to come dressed in period formal attire and would be served steak and fine wine; those who chose steerage would receive the appropriate meal for that class and have to scramble to find a seat. The party was held on a barge adorned as the *Titanic,* which was floating on the lake adjacent to my property. As each person arrived his or her photograph was taken in their period clothes with an old-fashioned box camera, ostensibly to be used for their "passport." Despite the obvious menu advantage of first class, an equal number of people believed that steerage might be more fun.

As the dinner ended and the band played on, the barge moved slowly offshore under a beautiful sky—until suddenly, and unexpectedly, it hit an iceberg. And then began sinking into the lake. This was not a facsimile sinking; the barge actually broke in half and was really going under. And, naturally, there were not enough lifeboats on the barge to save all of the passengers. Some of the passengers were able to find a seat in those lifeboats and made it safely, and completely dry, to shore. But others had to swim for it. It was quite dark and the water was cold. In fact, there were numerous scuba divers in the water prepared to help anyone who needed assistance, so no one was ever in danger. But people didn't know that, and most guests really got into their roles. In my mind I can still see the mayor standing stoically on the bow smoking a stogie and holding a glass of brandy as he went down with the ship. But the mayor's wife had less fun. She got in a lifeboat and almost made it to shore, but someone swamped it and ended up dunking her. She was clearly not happy about this, and while her husband and I subsequently became friends, I fear she held a grudge against me for a couple of years.

The period photographs were actually used in the "newspapers" reporting the event that were "published" as each person came ashore. Each paper was customized with a headline reporting that person's fate. Those people who made it ashore dry were reported to have "miraculously survived," while everyone who went in the water suffered a "tragic death" on the *Titanic.* For my passengers it was a far more memorable experience than simply attending a theme party, as they actually were forced to participate in this adventure.

That was also true, although in a very different way, for my party of illusions. Great stage magic is interactive; it uses some of our senses to fool our other senses. Even while watching an illusion, the audience participates by trying to figure out how it is done. They know that the woman in the box is not really being sawed in half, even when both halves are separated and seem to move independently, but the fun is in trying to figure out how the trick is performed. And that became the backstory to my event, Magic at the Manor, which I crafted with my collecting and magic best bud Brad Henderson.

Even the invitation, which resembled an ace of spades, was interactive. A nonsensical three-line poem on the back read, "Mangled Cats. The Man Soars. Can You Imagine?" It was seemingly meaningless, and intentionally a bit strange and dark. But when the recipients saw it was from me, they knew there was more to it than what was printed. People began talking to friends who had also received this invitation, trying to figure it out. Several noticed that at the very bottom of the card, where there might normally have been a copyright notice, in very small letters there was a website: www.gaffus.com. Aha!

People who went to that website discovered it consisted only of an equation, which read: "32°F < t < 451°". Now, most people know that 32°F is the temperature at which water freezes, while people who read the book *Fahrenheit 451* are aware that paper begins to burn at that temperature. The equation was giving visitors a clue to consider temperatures that were less than that at which paper burns but greater than the temperature at which water freezes.

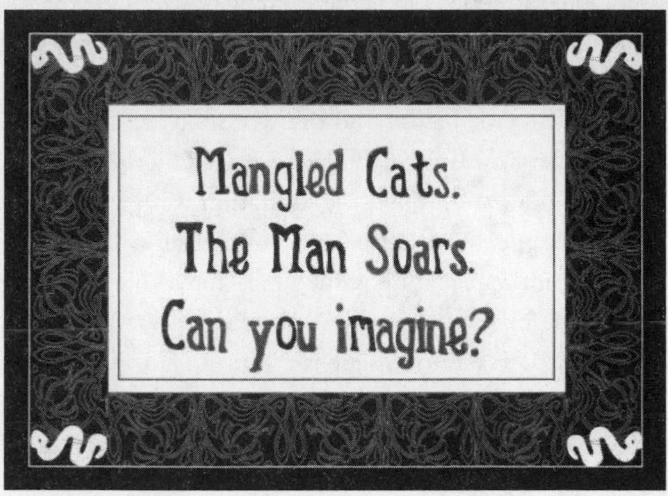

The magical invitation to Magic at the Manor.

While some attendees were busy trying to deboogle the meaning of the meaningless poem, the more clever people tried heating up their invitation, being careful not to burn it. And as they did, half of the poem faded away and the words "Mangled Cats. The Man Soars. Can You Imagine" were replaced by the web address magicatthemanor.com. When the invitation was put in a freezer, that phrase disappeared and the poem "magically" reappeared. It was printed with a special ink called FriXion, which is normally sold as erasable ink. Rubbing the ink with an eraser creates the heat necessary to make it "disappear," but it is never actually removed from the paper.

The website promised that at the end of the evening the secret behind the greatest stage illusion ever—cutting a woman in half—would be revealed. All the best stage magicians at some point in their careers have done this trick, and there are many different variations on it. Its inventor, a man named Alan Wakeling, never revealed the secret of how he did it. But as our grand finale, for the first time in a century, our magician was going to perform the trick as Wakeling had invented it and then demonstrate exactly how it was done!

It immediately became clear that some people did not want us to reveal this secret. In fact, when our guests arrived at the party, they were met by several angry protesters holding signs, people who had heard that we were going to reveal one of the great secrets of the world of magic and didn't like it at all. The local media had also heard about this and several TV reporters were there asking guests how they felt about the whole idea.

We had spent a lot of time and money producing this evening and, unfortunately, things started to go wrong right from the beginning. During the opening act sandbags fell from the ceiling and destroyed a large prop. As the evening proceeded, things continued to go wrong, and even some of our performers stepped out of their personas to complain about our plan. Eventually many people in the audience revealed their own feelings about our plan to divulge this long-held magic secret.

Truthfully, there was no such plan. There weren't any real protesters or media—we had hired them all. And we weren't revealing Wakeling's secret. The night turned out to be a mystery evening where the tricks were real, while all the controversy surrounding them wasn't. We did, however, saw a woman in half. We had a very famous magician who did an especially astonishing version of the trick, in which a woman was dissected within a few feet of the audience and the two halves of her body were carried on silver platters to opposite sides of the stage—at which point the upper half of the body wakes up and realizes the horror of the situation, and with her dying gasp interacts with a member of the audience.

Perhaps the most outrageously interactive event I've hosted was our *Mad Max Beyond Thunderdome* party, which was actually an exploration of physical boundaries. We had fire jugglers and fire walkers who taught our guests how to walk on hot coals, we had a suspension show in which performers put hooks in their bodies and were raised into the air, and we had a real tattooing booth and body piercing booth and invited guests to be tattooed or pierced. Several of my friends predicted that guests would be eager to learn how to walk on those burning coals but that no one was going to be idiotic enough to get a permanent tattoo.

I disagreed. I bet they will, I said, especially if we've got a talented tattoo artist.

Well, maybe a small tattoo, but they definitely will not get anything pierced.

I smiled, a smile that others might have described as cunning. Being tattooed or pierced requires commitment; these tattoos and piercings were going to be there when my guests woke up the next morning, and every morning after that.

We had two tattoo artists and one piercer, and they had the

longest lines throughout the night. Not surprisingly, one of the people in the piercing line was me. I had already decided I was going to have a piercing—it was my party and I'll pierce if I want to—I just hadn't decided what it would be. And truthfully, there is something somewhat unsettling about the term "creative piercing." To comfortably fit my personality, it had to be visible, subtle, and unique. Before continuing, pause for a moment and try to imagine a piercing that would fit those requirements.

Besides that.

Finally I decided to have my ear pierced—sideways. Often that piercing will go unnoticed, but on occasion I'll catch someone staring at it, trying to figure out what looks different. That gives me the opportunity to tell them the stories of these strange parties I've thrown.

I've approached all the interests of my life, from creating video games to traveling in space to throwing a theme party, with the same type of commitment I had when creating my original haunted houses: I had no desire to do a haunted house featuring an actor dressed as a goblin or an orc who jumped out and said boo! I wanted people to come over the rise of the hill, go around a corner, and come through trees to a different reality. I wanted to take people beyond everyday reality and the structure and the rules and the history of our world, to bring them—even for an instant—to a fantasy realm capable of evoking the emotions of real life. In my games and in my life, I challenge the assumptions of reality by creating moments of such awe and inspiration that it will feel like the rules of physics have changed.

Tech-Savvy Facebook Users, I Dare You!!!

www.takethislollipop.com

I DARE YOU

10 CREATE

The Magic of Science

The Magic of Science

I like to play with the rules of science when I can. I have a deep love for both science and magic and have spent many years learning as much as possible about both of them. I've found that people often confuse the two.

That confusion put me in a potentially embarrassing situation one evening. I usually carry a six-sided die with me with which I perform either a scientific act or a magic trick. I will place it on a table and let someone turn it to any number they choose while hiding it behind an opaque object so I can't possibly see their number. I claim that by reading their mind, I can correctly tell them the number. I then look deeply into their eyes and ask a few simple questions—then tell them what number is on top of that die. I can do this feat with uncanny accuracy.

One night I was performing this trick for my father and a group of other scientists and engineers at an Astronaut Hall of Fame event. My father, an inductee, laughed uproariously when I showed it to him . . . but then told me, "Whatever you do, do not to show it to Ed Mitchell!" Ed Mitchell is a celebrated astronaut, the sixth man to walk on the moon. But . . . as I would discover, he has also spent his lifetime investigating paranormal phenomena and claims to have successfully conducted ESP experiments from space with people on the ground. Ed Mitchell is the only astronaut I know who believes these things, and he takes them very seriously.

Unfortunately, just as my father finished warning me not to show this trick to Ed Mitchell, Ed walked up and noticed that I was holding a die in my hand and a crowd was gathered around me. "What are you doing, Richard?" he asked. I looked at my dad, who shook his head, implying, "Well, you kind of have to do it now."

I let Ed roll the die, then looked in his eyes and asked him my questions. And lo and behold, I got the number right. "That's great," he said, his face lighting up—and then he added, "But I know how you did it."

I doubted he could figure it out so quickly. "Really?" I said. I was curious to know what his speculation was going to be. "Okay, how did I do it?"

"You used quantum resonance," he said flatly.

"Quantum resonance?" I asked.

He nodded and said, "I have a friend who is very good at quantum resonance." This person, he explained, was capable of mind reading.

I didn't know what to do. It was an awful situation to be caught in: There were a group of very accomplished people standing around listening to this and I didn't want to be disrespectful to someone with Ed Mitchell's accomplishments by telling him, "No, it's just a magic trick. I didn't actually read your mind." But I also didn't want to tell him, "Wow, you got me, you're right, I'm using quantum resonance." So I decided to give him a broad hint, figuring he would understand what I was trying to tell him so we could both walk away happy. "That's possible, Ed. It might be that you're on to a method that I might be using, but I have to confess to you that I'm a bit of an amateur magician so I might have used some other method."

Ed didn't want to hear that, and said firmly, "No, no, believe me, I know how you did it."

I kept trying to give him clues, anything short of, "I'm cheating. It isn't real, it's a trick." But he was insistent that this was a demonstration of ESP. To this page I suspect he still believes I have that power.

I confess to you that this is definitely not proof of quantum resonance, it is a trick. Every magician, even an amateur like me, is often asked to explain how they performed a trick. When I'm asked, I always respond that I really can't do that. Sometimes people insist, promising that they won't tell anyone else. When that happens I lower my voice conspiratorially and ask, as many magicians do, "Can you really keep a secret?"

What would your answer to that question be? Without reading your mind, I am confident it would be the same as most people, who without fail will look at me and say sincerely, "I can."

To which I always respond, "Good. So can I."

The Science of Magic

The Science of Magic

It isn't always so easy to figure out the difference between science and magic, as basic science is often the trick behind great stage magic. I have a large collection of magical apparatus, for example, and among my possessions is an ordinary-looking wooden table. If you come to my house, chances are I'll show it to you. There is an object on the table, and I'll ask the smallest person in any group to pick it up, which they will do without any difficulty. After casting my spell I'll then ask the strongest member of the group to pick it up, and no matter how he—or sometimes she—huffs and puffs they will not be able to do so. Part of the fun is watching that person's growing frustration. A century ago this trick dazzled audiences in vaudeville theaters; a little child easily picked up a metal object sitting on the stage, barbells, for example—the same object that two or three large volunteers from the audience couldn't budge.

Science prevented them from picking it up. The "magic" is actually an electromagnet hidden under the table or beneath the stage; when surreptitiously turned on it exerts tremendous force, locking an object onto a surface, but when it is turned off that same object can easily be picked up and moved. In those days the power of an electromagnet wasn't widely known. That's science disguised as magic.

Science often looks a lot like magic: Let me show you this:

Before we had running water, people would shave over a bowl filled with water. I have built a table with a metal bowl in it, just like the type of bowl people used those many years ago. Inside this bowl I have put a viscous dark liquid. Since it looks a lot like blood, when I am doing this demonstration I often call it blood, to chilling effect. I'll begin my demonstration by asking a careful observer to pick up the bowl and examine it closely. Even closer, I allow that observer to stir it, shake it, do anything he or she wants until they are satisfied that there is absolutely nothing hidden inside. Then I ask all the observers to come closer to the bowl and peer into it. And then I beckon the creatures to arise from the great depths of this bowl.

Slowly, very slowly, from six different places in the bowl three-dimensional sea-urchin-like creatures rise out of the dark fluid. This is not a trick, it is very cool science, and you can try this at home.

There is a very uncommon material known as thermo-fluid, which will react to the presence of magnetic forces. What observers don't know is that I've hidden large electromagnets around the bowl that I can remotely power up or down to cause this reaction. It invariably results in shock and wonder.

An Illusion of Space

Space is the perfect meeting place of science and magic, because in space the rules of science are quite different from those on earth. I was able to test those properties myself when I traveled into space to live on the International Space Station (ISS). One unique property of space is the hard vacuum that exists outside the ISS, and another is the lack of gravity. When it became clear that I was going to fulfill my lifetime dream, I wanted to create the single greatest magic trick in history. Since I clearly was not going outside the space station, I wanted to do a trick using the properties of an environment with no gravitational pull. An obvious candidate would have been the classic floating woman trick in which someone seems to float unsuspended in midair, allowing the magician to actually pass a ring completely around their body. While that's an impressive illusion on earth, it's not so great in space, where everything floats all the time. After thinking it over for a while, I realized I needed professional assistance.

I sent a call out to as many magicians as I could reach, but got very few practical responses. One magician, for instance, suggested a trick that involved passing a card through the window of the ISS and watching the card float away in space. It was an interesting idea, but as setting it up required being outside, that wasn't going to be possible. I remembered David Copperfield's wonderful illusion in which he made the Statue of Liberty disappear. If I could have done a spacewalk, I could have made the ISS disappear, but the additional option of a spacewalk would have cost me $15 million, so the actual result of that trick would have been making $15 million disappear.

I finally decided to do a rising-card trick that was created by Robert Plants and suggested to me by my friend Brad Henderson, a trick in which I would discern through my magical powers a secretly selected card—then incredibly cause it to rise out of the deck and float. It's a great trick in which the techniques of stage magic and the properties of science combine to create a memorable illusion. I would pull one card from a deck of cards then return the remaining fifty-one cards to the box. Then I would open the box and the card I had already removed from the deck would rise out of the box; on earth that card would stop about halfway out of the box, but in space I could set it up so the card would rise and keep rising until it was out of the box and floating in the cabin. It was a rising-card trick that kept rising.

It almost didn't happen though, as NASA did not want me bringing a deck of cards onboard. They warned me: these are fifty-two small items, and if you lose them some will block specific air intakes that may overheat the fans and lower the air flow. They suggested I drill a hole in the deck and pull a string through it so none of the cards could be separated. Eventually I convinced them I had no intention of dealing cards or otherwise allowing the cards to float around the space station; rather, they would be held tightly together. Apparently I was so convincing that they permitted me to be independently responsible for my deck of cards. I brought my deck of cards onboard and performed this trick, which critics later praised as "out of this world!"

The Illusion of Magic

The seemingly magical illusions I present at my events are mystifying only because most people don't know or understand the science behind them. For example, I hosted a Mayan End of the World party at which we taught people how to walk on water. Walking on water has always been associated with faith; according to Matthew 14:28, Peter showed his faith by heeding Jesus's command to step out of his boat and walk on water. But it can be done pretty easily by mere mortals. This is a wonderful visual illusion—except it is absolutely real. The "trick" is to create what is called a non-Newtonian fluid. Think of being at the beach: if you pick up some muddy sand and pat it down, it becomes kind of mushy. Well, if you super-saturate water with any one of a number of different materials (in this case we used cornstarch), it suddenly has very different properties from normal water. This is another trick you can try at home: mix water and cornstarch in a roughly fifty-fifty solution. It's still a fluid, so pour it out onto a flat surface until you've got a puddle. If you really poke at it with your finger for a few seconds, it becomes stiff and rigid. But if you put your finger in the fluid, it will sink into it.

For the party we built a little pool about three feet deep, but we colored it dark blue so no one could tell its depth by looking at it. Then we invited people to walk across it. Many were dubious. But as long as they moved pretty quickly, they stayed on top of it (if they stopped, they sank). That's why this is known as walking on water, and not standing on water.

While I love magic, to me science is way cooler because it allows me to understand the way the world works.

In Search of Real Magic

People have often asked me if I believe in real magic. I tell them, "Just look around." We applaud stage magicians who turn a handkerchief into a dove, but a seed turning into a beautiful flower is true magic. Is there anything more magical than a caterpillar transforming into a butterfly? Indeed, nature sometimes gives us things that almost defy scientific explanation.

I remember the moment I held our first child, our daughter, Kinga, in my arms. I was overwhelmed with emotion. I had to sit down because I was so shaken I was afraid I might drop her. It was more than a week before I felt confident enough to hold her again. This may well have been the single most emotional moment of my life—so when I'm asked if I believe that real magic exists, the answer might well be, Have you held your child in your arms? The concept of two people creating a child is a very special kind of natural magic.

But when I'm asked if I believe, like Ed Mitchell, in things like ESP, I say no. I'm always a little careful to leave myself open to possibilities, so if I'm then asked if I know for certain that ESP doesn't exist, I would say of course not. Many things that exist are difficult to prove. DNA has existed since the beginning of life on earth but until we had the technology to view it, it was a theory, a belief. So it's quite possible other things exist that we just don't have the machinery to see yet.

I remember watching a TV program featuring esteemed primatologist Jane Goodall entitled *When Animals Talk*. The show focused on the impressive examples of communication between animals and humans. It was almost all factual. The final segment featured a woman who has trained generations of African parrots to do impressive things like recognize colors and shapes; when the parrot was told to pull out a green triangle or a red circle, it would do it. That's reasonable, I thought as I watched it, and I believe it. I think I still do. But at the end of the show, Jane Goodall went into another room, selected a photograph of an object from an envelope, and just thought about it. And the parrot, in another room, supposedly was able to name the image she was looking at. It was filmed in the same style as the rest of the show and seemed equally realistic. Except it was a trick. I know how this is done. Yet this program presented it as a great mystery of science. Oh no! I thought. A wonderful scientist had yet again been duped by a very good magician.

This isn't that unusual. The fact that a good stage magician can defeat significant scientific scrutiny poses a moral quandary: should they do it? I have a problem, for instance, with people who claim to be able to speak with animals and dead relatives. This is a very well-known, easily learned technique known as cold reading. It's "magic" being used to exploit deep emotions and, too often, to steal money, and obviously is unscrupulous. But did Uri Geller do anything wrong when he entertained people by claiming he could bend metal with his thoughts? He presented himself as an individual with unique psychic powers; that was his act and it made him seem very powerful.

What about ESP? For me, the difference between the science of magic and the magic of science is how useful it is as a tool. The laws of gravity are important because by understanding that large-mass objects tend to attract other objects you can use that knowledge as a tool. It's helpful when you're trying to make sense of the universe as it enables you to predict the outcome of some interaction. But if there actually is some form of ESP that is so vague it works only .0000001 percent of the time, it's not going to be terribly beneficial. As a tool it's about as useful as a hammer made from pasta. So it's not terribly important whether or not I believe it exists.

You were thinking the same thing, weren't you?

11

EXPLORE

The Discovery of the *Piña Colada*

On July 21, 1961, America's second astronaut, Gus Grissom, flew in space for fifteen minutes and thirty-seven seconds in a capsule named *Liberty Bell* 7. The mission was successful—but after landing in the Atlantic, *Liberty Bell* 7's hatch blew off prematurely and the capsule began filling with water. While Grissom was saved, the capsule sank more than fifteen thousand feet below the surface, far beyond NASA's capacity to recover it. In our space program Gus Grissom's *Liberty Bell* 7 was the only capsule that was not recovered.

When it happened it created quite a furor, as there were two possible explanations. One laid the blame on the engineers who built the capsule, the other blamed the astronaut operating it. A debate began that has never been settled: was the loss caused by mechanical failure or did Grissom err in blowing the hatch too soon?

People who knew Gus Grissom have continued to believe it had to be a malfunction because he was a very talented pilot and engineer and a cool-headed guy. In fact, even after this episode NASA had such confidence in him that he became the first man to fly in space twice and was selected to be the first man to walk on the moon. He never got to take that first giant step; instead he died tragically with fellow astronauts Ed White and Roger Chaffee when their *Apollo 1* capsule caught fire during a preflight test. On the other hand the engineers who investigated the loss of the capsule believe there is no possibility it could have been caused by a mechanical failure. The consensus among this great collection of engineers, scientists, and astronauts has always been that there simply isn't enough information on which to base a valid opinion. In other words, who knows?

In May 1999 the Discovery Channel funded an expedition to try to find Grissom's capsule and perhaps settle this debate. They hired colleagues of my business partner Mike McDowell to try to locate it using side-scan sonar. Part of the deal was that these people would keep the rights to anything else they found during this search. While searching north of the Grand Bahamas, they found the capsule; examining the sonar data it looks like a little teepee sticking out of the sand, but it was clearly the *Liberty Bell* 7. The Discovery Channel was eventually able to raise it, but found no evidence that made any difference in the ongoing debate.

But during the sonar search, the team also discovered several other intriguing objects, including what appeared to be a very old shipwreck. Two masts were sticking out of the sand-covered seabed and the outline of a vessel around those masts was also very clear. So a few of us decided to take our submarine down there and explore the wreck. Before making this trip we tried to identify this ship and determine when it had sunk, but we could find no record of one going down in that area. This truly was a mystery ship.

The wreckage was sitting on the sand about fifteen thousand feet deep, at least ten times deeper than any wooden ship that had ever been discovered, explored, or salvaged. It was three thousand feet deeper than the remains of the *Titanic,* and far deeper than could be reached with most submersibles. Like with the expedition to the *Titanic,* extremely specialized craft were required—and we had them.

We went down to the site in two subs, *Mir 1* and *Mir 2.* Once we were hauled over the side of the mother ship, it took a couple of hours to sink all the way to the bottom. In the Gulf the ocean floor is a vast sandy plain, much like an undersea desert. We slowly floated along just above the sandy bottom searching for the wreck. For a time we saw only bits of trash that had fallen off passing ships and then, strangely enough, golf balls. The seabed was littered with golf balls. I later learned that it is not uncommon to find driving ranges on cruise ships, so people drive golf balls into the ocean, presuming these to be harmless bits of trash, but in fact they collect on the bottom and serve as a reminder that civilization is never far away.

The water quality at that depth is quite good, but it is far too deep for there to be much light. The submarines carry a large number of very bright lights, and often travel together to maximize the light regions before them. Visibility is limited not by debris in the water column but, rather, by the reach of the lights.

We didn't go directly to the actual GPS coordinates; instead we tried to triangulate its location. As we reached the depth and motored over to the site, the first thing we saw was a single mast, with rigging tangled up on it, but not the ship itself. The sails and ropes led off in another direction and we followed that trail of ropes as they stretched across the seabed—and there it was, the hull of a wooden ship sitting upright on the bottom.

There is little I find more compelling than a previously undiscovered artifact of the past. It is an extraordinary feeling knowing that you're the first person to be looking at an object in several hundred years. It's what every explorer and adventurer lives for. As we examined the wreck, in some ways we were like the engineers who were examining *Liberty Bell 7* and trying to understand the story it told.

It was obvious why the ship had sunk: there was a four-foot by four-foot hole where the entire central mast had been ripped out. The mast was still attached to the hull by ropes and it lay about a hundred yards away. This ship had sunk quickly in a big storm. It is inconceivable that anyone could have survived in the near-freezing water. We had no doubt men had died on this ship. This was their grave and we were respectful of that.

The cold water at that depth helps preserve wood; a lot of this ship had deteriorated, but two of the masts were still upright. The upper decks had collapsed, exposing the decks below, so we could look into the ship. The front half was a cargo hold and we saw that it was still half-filled with coconuts and rum bottles that had never reached their destination. So we immediately named this ship the *Piña Colada*.

We theorized that it might actually have been a slave runner, a ship that took slaves in one direction and carried legitimate cargo on its return trip. As we drifted over the hold, we saw a variety of items floating upward, but still tethered to the ship by barely visible debris, looking something like odd-shaped balloons. There were pieces of silk cloth, which we were able to grasp with our robotic arm and eventually bring to the surface. But most oddly there were books, which seemed to be held to the wreck by a thread. They were fanned open, moving slightly in the current as if being read by the ghosts. As we got closer we could see they had begun to deteriorate. Using our robotic arm we pulled out a very fine net large enough to hold an entire book. Very, very deliberately we moved that net into position around one book, then gently moved it forward until the book was inside it. But as the net touched the book, it instantly dissolved into a mist and dissipated, a cloud of thoughts lost forever.

As a lover of history, I found this journey very special. The wreck was an open window into the past. In the captain's quarters we found some pots and jugs, some pistols, cups, plates, and bowls, and finally, a treasure chest. Mike McDowell actually discovered it, but all of us shared in the elation. We had seen a shape and identified it as a likely sunken ship, then gone down to discover the deepest wooden wreck ever found, then searched that ship and discovered in the captain's quarters a chest filled with silver and gold coins. This was the culmination of childhood fantasies.

The chest was about the size of a footlocker. We picked it up gently and brought it back with us. Much of the chest had disintegrated, but we found a brick of solidified silver coins, and inside the brick was a little gold box about the size of a cigarette lighter. The archeologist we'd brought down to the wreck with us put that small box into a water bath to prevent it from deteriorating and opened it slowly to reveal more than a dozen gold coins wrapped in newsprint. There is no reasonable explanation for why that newspaper page survived mostly intact while the books dissolved.

We were able to tease open a piece of it and from that, as well as the dates on the coins, we were able to determine a general sense of the history of this wreck. The paper was in English and it was plain enough for any of us to read. It described plantation owners and prices of goods throughout the Caribbean—and the need to continue certain kinds of trade, including slaves. So our supposition about the period and the history of the ship was confirmed to a great degree, although we were never able to identify the ship or find a definitive history of it.

12 CREATE

The End of the Beginning of the World

I have been credited, somewhat inaccurately, with inventing the term and category of games known as MMORPGs, massively multiplayer online role-playing games, because *Ultima Online* was the first to become a major success. Before *Ultima Online,* every online game was played by a solo player. It was the player against the game, so everyone could be the hero who defeated the bad guy and won the game. And it didn't make any difference that their best friend also believed they were the hero who'd won the same game. People would talk to each other about where they were on the quest and trade tips about how to get past a certain point. No one felt less special because a friend was ahead of them, because they couldn't see the friend unlocking a door in front of them or beating them across the finish line.

But in fact, long before the existence of the Internet, people were linking computers to communicate with each other. They got together through dial-up services like the original America Online, and they paid for access by the hour or even by the minute. A decade before we produced *Ultima Online,* we were already dreaming of producing a game that allowed many people to play in the world of Britannia at the same time and interact with each other.

We were not the only people with that dream. Since the availability of computers, some people were producing text MUDs, or multi-user dungeons. These were text-based games, although a few of them had very simple graphics. These multiplayer role-playing games were quite simple, and the level of player interaction was about as basic as it could be. Players could chat; one could write, "Hi, Richard," while another could write a command like, "Go north." And on the screen, sure enough, a figure would move to the north. But these games had user bases in the few hundreds or low thousands, and were not competitive with "mainstream" games. These games never really appealed to me nor did they inspire me beyond seeing the potential they offered for something much more exciting.

We watched this segment of the gaming industry very carefully for at least a decade. We'd meet regularly with the companies making the best dial-up games to discuss producing a multiplayer *Ultima.* But with the fee structure that existed at the time, we couldn't figure out how to make a business out of it. These games were expensive to play; they required a subscription to a dial-up service and the game itself generally would charge as much as a few dollars an hour to play. It quickly would cost considerably more than a boxed game purchased in a store.

The availability of the Internet, which allowed people to be online for extended periods of time without being charged by the minute or hour, completely changed the economic structure. Suddenly, a million people could be online at the same time! Two million! A hundred million! The potential was almost incalculable. We knew it was the right time, investment costs notwithstanding. So we set out to create a game that would appeal not just to the few who played MUDs, but to the millions of people who were playing any type of computer game.

While the original *Ultima Online* was not the first online role-playing game, it was the first widely successful massively multiplayer game, and its success proved that a huge market for multiplayer online games existed. As a result, I was identified as the person responsible for bringing MMORPGs into existence, and have even been called "the father of online gaming." I've never described myself that way, but it sort of caught on—much to the chagrin of those hearties who had been making MUDs for many years.

A few years after the successful launch of *Ultima Online,* for example, I was asked to speak at a computer game developers' conference in Austin. I was in the green room picking up my credentials when a man I'd never met in person, Mark Jacobs, approached me. At that moment all I knew about Mark was that he was the creator of a very good MMO called *Dark Ages of Camelot,* which had been published a few years after *Ultima Online* and had done very well. As I later found out, he had also made some of those dial-up games, those text MUDs, which predated any online game I had done. "Richard, Richard Garriott!" he said.

It's not unusual for people I've never met before to approach me at events and I usually enjoy the conversations that follow. So I smiled at him and acknowledged that yes, I was Richard Garriott. "What's up?" I said.

"This is outrageous," he said far too aggressively.

"What's that?" I asked.

He held up the brochure for the event. "Have you seen this? Have you seen what they've written about you?"

"Uh, no, actually I haven't," I said. "In fact, I just got here. I haven't even gotten my badges. What's this all about?" Apparently I was in the middle of an argument and I had no idea what it was about.

He introduced himself, then said, "It calls you 'the father of online games' in here. That's ridiculous. I demand that you retract that." He said it quite loudly, and all the other developers and businesspeople in the room stopped whatever they were doing and began watching us.

I was with my friend, our corporate publicist David Swofford, who tried as politely as possible to defuse the situation. I said as calmly as I could, "It's not up to me to retract it. I didn't write this, I haven't even seen it. I just got here."

That made him even angrier. "You can't claim that title. It isn't true. There were lots of online games before you."

As far as I knew I hadn't claimed anything, much less a title. Of course there were others before mine, I thought to myself. "I had nothing to do with this. The show produced the brochure. You should go talk to them."

But he had no interest in talking to anyone else. He continued ranting, "You need to publicly deny this right here in front of your peers."

"Look," I said. "I still don't have the slightest idea what you're upset about. But I didn't write this and, truthfully, who cares?" In fact, it made no difference to me.

Well, he told me who cared. As he continued berating me, David Swofford wisely grabbed my arm and basically rushed me out of the room. But that was not the end of it. Mark and I both served on several panels, although fortunately never on the same one. And at the end of each panel he stood up and said something like, "I just want to bring it to everybody's attention that this booklet says Richard Garriott is considered 'the father of online games.' Whoever wrote that should be chastised. I think it's an outrage that he should get this credit when in fact there were a lot of us, including myself, who . . ."

This confrontation had the exact opposite effect that he'd intended. After Mark Jacobs decided he was my nemesis, I began receiving notes, stickers, and even handmade certificates congratulating me on being "the father of online games." Before he started this debate, I don't know that anybody cared at all about this, or even took it very seriously; I certainly didn't know about it—but after he made it an issue, my place in the history of online games was solidified. People thought I must be the father of online games if someone was making such a big deal out of complaining that I wasn't.

In the ensuing years I've had several far more pleasant encounters with Mark Jacobs, who turned out to be a good and very competent guy. I suspect (until reading this) he may have completely forgotten about that incident. And ironically, after EA shut down Origin, they needed someone to manage *Ultima Online,* and for a period of time Mark Jacobs became the head of *Ultima Online.*

I like Mark, he does great work, and I wish him the very best. He beat me by decades into online games. But *UO* still wins the crown for the first true MASSIVELY multiplayer game. So I will happily accept the honor of being known as the creator of MMORPGs and the father of online games.

Maybe *UO* wasn't the first, but it was completely different from any game that had been created before it. It was obvious to me that when we put many thousands of people in the same online world, we couldn't write the same old story in which everyone was victorious. Not everyone could defeat the one evil lord. Not everyone could drop the one ring in the fire of Mordor. This was no longer a story that froze in place when a player turned the game off for the night and resumed when it was turned back on; rather, it was a world that existed twenty-four hours a day, seven days a week, fifty-two weeks a year whether you were playing the game or not, and more than that, it was continually changing. *Ultima Online* has no linear story. There's no quest line that drives you forward. It isn't about winning or losing, it is a world in which people can live a virtual life. Players make up their own goals and objectives within the greater context of the world we place around them.

In fact, it was so different that initially EA had no interest in developing it. Solo games were selling millions of copies, while the rudimentary MUDs were selling a few thousand at most. My team and I explained to EA, for whom we were then working, that this new concept, massively multiplayer online gaming, was the wave of the future and they needed to be prepared to ride it.

No interest. They pointed out that there was no data to support my belief that anyone would want to play an online game. They said, "Nice idea, Richard, but there's never been a successful game of that type. That's just not the way people want to play games. And we're not interested in getting in that business."

EA had a very formal process that its developers had to go through before the company would even consider funding a game. To make a case for funding, we had to put together a huge set of documents that demonstrated why it would be a success. The sales department did an analysis based upon the sales of similar games in the past. So while they were comparing the potential of *Multima* to the small numbers of MUDs that were selling at that time, we were trying to get them to make an investment in the future. It was easy to understand why there was so little enthusiasm at EA for this type of game: since no one had ever earned a penny selling the type of Internet game we wanted to create, the sales projection was essentially zero.

EA also pointed to the widespread belief throughout the industry that medieval fantasy had lost its luster. While *The Lord of the Rings* had been popular years earlier, the movies had not yet come out, there were no Harry Potter books, there were no good medieval TV shows. Instead, it was pointed out to us, sci-fi fantasy was what was selling. Everybody wanted some version of *Star Trek* or *Star Wars*.

While the sales department contrasted fantasy and sci-fi, I have always believed they are effectively the same. I can't think of a fantasy story that couldn't be told in a science-fiction setting, and the reverse is true. I think that most of the people who love *The Lord of the Rings* also love *Star Wars*. But the sales department pointed out to us that none of EA's competitors were producing any fantasy games without sci-fi elements. Someone in the sales department essentially told me, "Richard, we're not going to give you good sales projections if you keep insisting on doing medieval fantasy games. No one wants to run around in a medieval world wearing tights like Robin Hood. They want to wear dark glasses and trench coats and shoot guns like *Men in Black*. So if you want a good projection, you have to drop this no longer popular medieval fantasy stuff and convert your storytelling to science fiction."

Of course I wouldn't do that. I told them flat-out that they were wrong, that the only reason fantasy wasn't selling was because no one had made anything good in that genre for the past few years. It was a lack of quality, not the popularity of the genre. I made the comparison to *King Kong*. Several schlocky big monster movies came out and didn't do well, so people decided the *King Kong* franchise was not a valuable intellectual property. Then Peter Jackson made a great movie and suddenly *King Kong* was hot again.

But EA wouldn't budge.

There's a lot about running a business that I don't know or that I'm not very good at, but when I have an idea I believe in, I am tenacious. Six months later we went back and told them, "Hey, take a look at that Internet thing, it's really starting to catch on. Now is the time to get moving."

For the second time we were told no.

Six months after that we went back again. I literally would not take no for an answer. This time I went into CEO Larry Probst's office and told him, "Look, we spend five million or more on any game we develop. Give me $250,000 to prove to you this is viable." Larry was a fine businessman who believed the numbers told the story; I tried to convince him that there were times numbers didn't paint a complete picture. He truly was not interested, but I wouldn't leave the office until he agreed. Finally, he agreed to sign a note I'd already written out that said basically, "I hereby grant Richard the ability to go over budget by $250,000 in pursuit of this *Multima* thing that he wants to do."

EA gave us the minimal funding, but no other support. We were last on the list to get any resource. We only got to see the resumes of potential hires after every other project had passed on them. There was a reason for that beyond EA's MMORPG skepticism: When *Ultima VIII* was published, it hadn't reached expectations. Instead, our game *Wing Commander* had become the most prominent money earner for the Origin division of the company. Because *Wing Commander* was doing so well, and it needed resources, when resumes came into Human Resources that team had first pick. And second pick. So our team became the ragtag group of rejects. We were like the Bad News Bears of gaming.

To make matters worse, EA wouldn't give us proper office space. The building was being refurbished and every other team was given new space as it was completed. The team my long-time creative partner Starr Long and I built was literally put out in the hallway; our desks were in the corridor facing floor-to-ceiling plastic sheets. Behind those plastic curtains the walls were being demolished and rebuilt. We were working in the middle of a construction site, and we had to learn to live with the bangs and screeching of hammering and sawing and way too much dust for computers. Everything came together to make us feel unimportant and unwanted.

In 1996, while we were making *Ultima Online,* another potentially "first" massively multiplayer game, *Meridian 59* was published. Visually it was nicely done and it was produced by a competent developer; it looked good . . . but it didn't do very well. I felt it was just a little shy of the critical mass necessary to take off. It wasn't a deep enough, rich enough virtual world to meet the next-step expectations. At the time, it had sold only about thirty thousand units. That was very scary for us. It certainly didn't help us make our case to EA executives, who were already skeptical and used that as a data point to reinforce their Richard is an idiot and don't waste our money listening to him argument. As it turns out, *Meridian 59* developed a loyal following and is, notably, one of the survivors from that early era. Its fans and original developers still keep it going on an open source basis.

Still, for the $250,000 we managed to build a nice little prototype. We then put up one of EA's very first websites to introduce the game. Basically we introduced ourselves: Hello, we're the *Ultima* team, and we've spent $250,000 making a prototype for a new game and we need people, a lot of people, to help us test it. "We can't download it to you," we wrote, "because it's too big. We need to send you the game on a CD, which is going to be expensive for us to manufacture and mail. So if you want to volunteer to be a beta tester and help us develop this game, please send us $5."

This was a marketing strategy no one had tried before: ask people to pay for the privilege of volunteering. We were hoping to find enough people to populate our small world so we could see what worked and what needed improvement, but we didn't know what kind of response to expect. The EA marketing team had projected lifetime sales for *Ultima Online* at about thirty thousand units—which they thought was wildly optimistic. They made that estimate based on the fact that the largest-selling MUD in history had sold about fifteen thousand units—and they doubled it. Of course, in this industry even thirty thousand units is irrelevant in a financial sense.

This wasn't the launch of a game. There was no advertising. This was a search for paying volunteers. The only publicity was its website. We put it on the Internet and held our collective breath—within a week or so fifty thousand people had signed up to pay $5 for the disc!

Finally, EA got it. *Ultima Online* instantly became the most important thing happening in the EA world. The floodgates opened; we got all the money and the development people we needed. After practically ignoring us, EA now decided we needed considerably more management oversight, so we inherited an entire level of EA management people. We were making things up as we went along, but these executives sensed that something important was happening and wanted to have a stake in it. While having money made the situation much better, the managers who came with it often made it worse. We were just trying to keep our heads down and make a game, but other people started playing company politics. Everybody wanted answers to questions we hadn't yet decided on: when could we ship the game, how much money did we need, who would be in charge after the game was online?

In the past, once we'd shipped a game, it was done, it was out. It was time to focus on "what's next?" This was completely different; we were going to be running a live service. It was like producing a TV show and having to continue producing episodes or your audience would go away, except it wasn't really like that at all. Or it was like being in an airport control tower without ever being able to leave and landing an endless string of planes, except it wasn't really like that either. There was no one in our group with this level of experience. We didn't even think about this when we were working; we were too busy just running around. But during beta testing we realized we needed someone to "moderate" the world we were creating. We needed a "community manager."

In fact, we hired what may have been the very first person in the industry to have the official role of "online community manager." We hadn't even created a title for that position when we hired the first one. My brother Robert tried to figure out what this person would do: "We know they aren't going to make art, they aren't going to write code, they aren't going to build maps. You're telling me they're going to be the person who sits on the Internet and talks with your players? Really? You want to hire someone to talk to players and the game's not even generating revenue yet? That sounds like idiocy."

After creating the position of community manager we tried to explain what it was. An online world has no political infrastructure, so the people in it have no means of communicating with whoever is in charge and no clear voice to represent them. We needed a way for them to register their complaints, to express their needs, and to offer suggestions as we moved forward. We also had to find a way to speak to them. At that time there was no website or other place where we could contact them. What happened next was amazing. The players and the company combined to invent many of the same solutions used in the real world, meaning that in addition to the people in the community employed by the company, the community self-organized and selected its own leaders. And at the center of all of this was our community manager.

We built *UO* one piece at a time. The first prototype that showed any aspect of multiplayer interaction was pretty simple: It was a grassy plain where I could move one figure and another member of the team could move their player—and we could see each other moving and we could talk to each other through cartoon-style text balloons. Then, naturally, we gave them simple weapons so they could club each other.

That fun turned to pure terror the first time we let in other players during beta testing. When I describe this moment, I ask people to imagine the city of Austin, which has a cozy population of just over one million. It's a big small city. Imagine that you invite thirty of your best friends to a dinner party and everybody sits down and over the course of a year those thirty people, with no input from anyone else, create an entire new city of Austin. Not just the architecture, but every aspect of it, right down to the candy bars on the shelves of the supermarket. Everything in that city, from the height of its tallest building to the price of a candy bar, has been created by one of those thirty people. Somebody decided where every pothole, every road, every fence would be. They decided how much power each building would need to operate, and then breaker boxes had to be installed to fulfill those needs.

The result is a beautifully pristine, completely empty city. All the lights shine brightly. When those thirty people explore their creation, everything works beautifully. They can go into any building and walk around, buy a candy bar, or cross a street without incident. There are no cars moving yet, but the roads are there. It is perfect.

Then they open the gates and welcome people into this city. One million people show up and start using the facilities. What is the probability that the power grid will stay online? Zero. What is the probability that the sewers won't back up? Zero. What is the probability that people will think the taxation rates are fair or that the economy will stay in balance? Right, zero. And what is the probability that when someone discovers their toilet doesn't flush, the lights don't work, and the garbage isn't being collected that they aren't going to be upset and try to find someone to complain to? One big fat zero. And when they get frustrated because they are trapped in a city that isn't working, and there is no one to tell them it's all being fixed, they get disruptive.

That's what happened when we invited people into our city for the first time. It actually happened twice: first, when we invited our beta testers into the city then again later when we went online for real. Neither time were we adequately prepared for the number of people who wanted to play the game or the widely varying ways in which it would stress unimagined aspects of the controlling systems of the game.

We had no expectations because we had no history to look at. The biggest complaint was that with so many people online at the same time, our servers were not sufficient to sustain their game play. We knew it and we were working on it, but apparently not quickly enough. There were other problems as well. Players might go down into the dungeons to unlock secret doors and discover treasures, only to find all the doors open, and the treasure chests ransacked. There were also limits to the number of people who could congregate in any one place. The more people there were, the more slowly the game would respond. It was inevitable that from time to time the servers would overload and the whole game would just come to a crashing halt. There were also countless times when the game's response to player activity would be inadequate or downright wrong.

The system was broken in so many ways that you would think players would eventually get tired of all the problems and pack up and go home, but that's not what happened. People clearly saw the potential; they desperately wanted it to work, so instead of walking away they registered their unhappiness in really clever ways, culminating in an act of civil disobedience that was impossible to ignore.

This was a remarkable social event and confirmed the viability of the game. These people knew each other only through their avatars, yet they were able to communicate successfully and stage a massive demonstration. It was pretty funny too.

Several thousand players simultaneously went to Lord British's castle and stormed the gates, which were locked. We had to help the protesters by opening them. Once we opened the gates, they flooded the building with bodies. People were packed in from wall to wall, from ceiling to floor. They knew that we had provided alcoholic beverages for the game; if you had one drink you would go hic, hic, hic; if you had two drinks you would respond hiccup, hiccup, hiccup; and if you had three or more drinks your character would get sick, double over, and vomit on the floor.

That was a small aspect of the game we had enjoyed developing: should you vomit after three drinks or four?

While the only limit on how much a character could drink was that it affected the character's performance, we did have a profanity filter that blocked characters from using certain words, figuring moms weren't going to like their kids hearing the colorful variety of language that some players were inevitably going to want to use. If someone tried to type in one of the more colorful words, the game automatically replaced it with symbols. So naturally players began figuring out ways to circumvent the filter, replacing the *s* in certain words with a dollar sign, for example. Hundreds of players crammed into the castle, took off their clothes, and started drinking and cursing. We quickly had a screaming naked mob vomiting all over the floor.

We got the point. People took endless screen shots, the press covered it, and, naturally, our servers crashed. The people playing took the entire system offline, very effectively making their point. Our problem was that not only were we more successful than EA had projected, we were wildly more successful than even we had imagined. Instead of thirty thousand units, we were going to sell several hundred thousand units, maybe more, maybe a million. We had designed servers that would allow thousands of people to be connected simultaneously, but we were completely unprepared for the game's success. At peak times rarely were more than 10 percent of subscribers logged on, so we believed we had sufficient servers to keep our world operating. But suddenly we had millions of subscribers, which meant that 10 percent was more than the total we had anticipated. Our servers couldn't handle it. This is commonly referred to as "the m-squared problem." Every player had to have access to every word said by every other player, a complex issue that required an enormous number of servers.

For a lot of technical reasons, we couldn't just make the world bigger. The only solution was to create a second set of servers and divide players between these worlds. One player logged on to copy A, while someone else was on copy B of the world or copy C or D. Nothing about these worlds was connected in any shape or form. I was very unhappy that we had to do this. My vision of a metaverse was not going to come to fruition, exactly. To me, this was a defeat. I knew that events in one world were not going to match events in another world, and the worlds would evolve independently and end up being very different. One world might have all the happy players, while another world would be where the dissidents lived. To me, this violated the basic concept of creating a fictional reality. I didn't want anyone thinking they weren't in the real world, but rather that they had been shunted off to a copy of the real world. In order to create a singular reality, I needed to find a fiction that tied everything back together.

For inspiration I went all the way back to *Ultima I,* which was the story of the evil Mondain the Immortal Wizard, who became immortal by possessing the gem of immortality. In order to kill him, the gem of immortality first had to be destroyed. So I wrote a story of how when the gem of immortality was broken, it took a snapshot of the world as it was at that moment and created shards of the gem. The reality of the world was reflected within each shard and those shards were scattered through the universe, to finally be found in *Ultima Online*. Every shard was an exact duplication of the world as it was when Mondain was finally killed, and in theory, if all those shards were collected and brought back to Lord British, he could reunify the world. So even though players might be in a disparate world, a shard of reality, it was possible that someday the world would be unified.

I never expected the word *shard* to become a common term in the online gaming world and even be applied to database server architecture. Banks, for example, use it to describe their networking capabilities. I was talking with someone from another company one day and he explained, "We've broken up our world into five shards and . . ." I stopped him and asked why he'd used that word and he explained that it was just the term used to describe a copy of the world.

"It's not," I told him, "that's not it at all. The word *shard* comes from Mondain the evil wizard's gem that was destroyed, and these shards had to be unified to bring the worlds back together."

"Oh," he said, "okay."

Eventually we were able to get twenty copies of our world running at the same time. But our challenges were only just beginning.

The Long Good-bye

It took us about a year to work out the problems that surfaced during the beta testing of *Ultima Online.* We were inventing solutions on the run, just trying to keep the world functioning. Several times during the beta phase we made sweeping changes as we came to understand the needs and quirks of our world. We tweaked how easy it was to get gold, and to advance in levels. Each time we made a change, we informed our beta testers that we had to reset the servers and erase all the player data, meaning their avatar had to start over. But we really didn't want to have to do that after the game went live. So the transition from the beta phase to live would be the last time the servers would ever be wiped. Everything about the world up to that moment would be erased, and when the servers came online the next day, the game would begin for real and everyone's avatars would theoretically live forever. From that moment on, whatever role people played, whatever fortune a player accumulated or property they owned, it would remain with them just as it would in real life.

We informed all of our testers of when we were going to wipe the servers clean for the last time. It was a major event, although even we had no idea that it would become one of the seminal events in the history of gaming. In fact, there isn't a list of the most important moments in game development that doesn't include it in the top ten, and usually in the top two or three. It was the night my avatar, Lord British, was assassinated.

This was the end of the world for the beta testers. Their homes, their businesses, their possessions would all be gone. At precisely midnight the servers would go down, they would stay down a few hours, and then the new and permanent world would begin. It was a monumental event for all of us, and our players wanted to be there to celebrate the end of the beginning. In addition to the players, every member of our staff was also online. All the senior management, the customer service department, and the designers were spread throughout our building, most of us in our offices, playing the game with the thousands and thousands of testers around the globe. I was with project director Starr Long, who was playing his character, Blackthorn.

To mark the occasion we had announced that Lord British and Blackthorn would go from town to town giving speeches thanking everyone and saying good-bye. We had to travel because if all the players tried to get into the same city at the same time, the game would bog down. Star and I teleported from city to city with many of our employees. It was kind of a king's entourage. We broadcast it on our global "network" so whichever town we were in, all the players could see it, as if we had a global megaphone. In every city there were hundreds of characters lined up, often very formally, to hear the creators make these final pronouncements. Sometimes people responded in very funny ways: When Star and I arrived in the town of Moon Glow, for example, we found the players lined up facing away from us. And then in unison all of the characters took off their pants and bowed. We were mooned in Moon Glow.

In fact, in each place Lord British went to make a farewell speech, people were finding fun and clever ways to celebrate this occasion. And then, just before midnight, just before the end of the world, we reached the city of Trinsic. Trinsic is protected by a high wall, and in the center of the city there is a large, open square. Much of the population of the city had gathered there to hear our final words in the final minutes of the final beta test days of *Ultima Online.*

At this point it's important to note that my avatar, Lord British, was immortal. Being immortal was absolutely essential because, as we had learned in the solo-player *Ultima*s, just about everyone's favorite activity was trying to kill the non-player characters, especially Lord British. Having to constantly fight for my life, escape assassination attempts, and avoid traps might be fun for other players, but not for me. By making Lord British immortal I could focus on building the world. And besides that, being immortal is cool. Who wouldn't want to be immortal if they had the option? Every time we did a wipe of the entire game, I had to make a new version of me, and when I did I re-created my special character, who looked like me and wore a recognizable outfit. I also set all my attributes—my strength, my intelligence, and everything else—at the peak of what the game could hold. Basically, I couldn't be harmed, and if I hit anybody with even my little finger they would be killed instantly. I even had a special flag called immortality that I could trigger, so that even under the worst conditions there was no way I could die.

It's good to be Lord British.

To maintain the peace we gave that same immortality flag to the guards in cities so that no one could overwhelm the police. In the waning moments of the beta version, both Lord British and Blackthorn appeared on the wall overlooking Trinsic's town square. As we began our speech, an avatar cast a fire-field spell up onto the parapet, setting the area directly around us on fire. Of course this was something we had seen many times before. During the early days of beta testing there had been countless attempts to assassinate Lord British and every one had failed—because he is immortal!

And so I treated this puny fireball as just another effort to be ignored. Ah, those mortal fools, believing they could kill the mighty Lord British. I took a couple of steps back to avoid the fire, but that meant I could not continue my speech; I could neither see nor be seen. Being immortal I had nothing to fear from this feeble attempt, so I decided to step through the fire. I enjoyed the thought of showing them my power by stepping out of the flames. So I took one step forward into the fire—and fell over.

Dead.

I was stunned. Oh. My. God. I couldn't believe it. Dead? Lord British?

As we later discovered, during the last server reboot several months earlier, I had been in such a hurry to create the character that I had forgotten to set my immortal flag. I hadn't noticed it because I hadn't been challenged; I had lived for months without any problems until we were minutes away from shutting down the servers. Suddenly, I was dead.

There was nothing I could do. When you fall over dead in *UO,* you can't speak, so I couldn't ask for help. And I couldn't resurrect myself. There were only minutes left in the game and my character lay dead in a storm of fire.

I was dumbfounded. It took a few seconds before anyone else actually realized what had happened. Then my staff set to work. While with today's technology it is pretty easy to do a quick query on the database to find out who performed what function, as *Ultima Online* was pioneering the development of this type of game, it took us days of going through our records to figure out exactly what had happened.

But at that moment that was not the staff's primary concern. They knew they had only five minutes before there would be an automatic shutdown of the entire service. This wasn't enough time to restore order to the proceedings. They weren't sure I would come back to life before then, and couldn't figure out which of the evildoers in this courtyard had killed Lord British. So there really was only one thing to do: kill them all.

It's amazing how quickly the cloak of civilization can disappear. The word spread verbally throughout the office: Let us unleash hell! My staff summoned demons and devils and dragons and all of the nightmarish creatures of the game and they cast spells and created dark clouds and lightning that struck and killed people. The gamemasters have special powers, and once they realized I had been killed, they were able to almost instantly resurrect Lord British. And I gleefully joined in the revelry; kill me, will you! Be gone, mortals! It was a slaughter of the thousands of players in the courtyard.

It definitely was not the noble ending we had intended.

And while some players enjoyed the spontaneity of this event, others were saddened or hurt by it. When most characters die they turn into a ghost and are transported to a distant place on the map. Then they have to go find their body. So the cost of being killed is a temporary existence as a ghost. In the last three minutes of these characters' existence, they suddenly found themselves alone, deep in the woods, unable to speak or interact with anyone else. But the net result of this mass killing in retaliation for the assassination of Lord British was that not only were all of these innocent people slaughtered, they were also cast out of the presence of the creators at the final moment. As the final seconds trickled down, they desperately tried to get back but most often failed. The fact that all of us, the creators and the players, were able to turn the last few moments of the beta test into this completely unplanned and even unimagined chaos was proof that we had built something unique, a platform that would allow players to do pretty much whatever they pleased, and it was about to take on a life—and many deaths—of its own.

We eventually proved that the person who killed me in those final hours was a player known in the game as Rainz. He lives in infamy to this day. And I still have yet to get my revenge!

13 CREATE

Nothing for Money

When we created *Ultima Online* it never occurred to us that our virtual world would make an impact on the real-world economy. No one anticipated that the game might enable students to work their way through college, or allow criminals and drug cartels to launder money, or lead to the creation of Chinese gold-farming businesses. We had no idea we were creating a real-money economy that would generate billions of dollars in revenue and would obscure the lines between the virtual world and reality. We just thought we were making a game.

Our business model was quite simple: people would buy the game for the retail price and install it, and then they would pay a monthly fee to play it. When we created the game, we planned and built all the elements of an in-game economy into it. Players could earn virtual gold in a variety of ways and use what they earned to purchase everything from a good virtual meal to a virtual house. To get the economy started, we sold some of the virtual items for modest amounts of virtual gold. It was simply a way of starting the economy. But we didn't offer anything for sale for "real" money or put real-money values on any items. We didn't establish an exchange rate either, assuming that as supply and demand took over, the balance of virtual gold to virtual items would find an equilibrium, which is exactly what happened—and then some!

Almost immediately people began to covet virtual items like property and magic swords but were not willing to put in the time to earn the virtual gold needed to buy them. Rather than "supply and demand," the way to describe what happened is "time is money." Instead of spending months playing the game to earn the virtual objects they wanted, players began making side deals, often through places like eBay, buying virtual assets for real U.S. dollars. So those people who were willing or able to put in the time necessary to obtain these items discovered they could sell them in the real world for real money. The accumulation of wealth and the discovery of rare and magical items were outsourced to places like China, which had easy access to computers and low-wage employees who could sit around playing games endlessly to accumulate the virtual wealth to be sold. This "gold-farming," as it has become known, became the standard for all massively multiplayer games.

I remember the first time I found out that an object from our game was being sold on eBay. It was only two or three months after the game became playable. A rare and magical sword was for sale for $100 and we were all flabbergasted. First of all, we wondered about the logistics. These sales require the trust that a seller will pony up after being paid. And the concept that people would pay real money, sometimes a lot of it, for virtual goods seemed bizarre. We weren't prepared for it and we didn't know what to do about it. In some ways we still don't.

The obvious question everybody asks, the first question I would have asked, is why anyone would pay real money for nothing. But for some people it actually makes a lot of economic sense. We had assumed that people would spend as many hours as necessary playing the game to earn virtual gold and then use that virtual money to buy virtual items, a new sword or a plot of land or even a good meal at the local tavern, for example. But for many people, it just wasn't worth it. A player could spend two weeks in the game mining monsters and collecting treasure before they had enough gold to purchase a big sword. Or they could pay someone to do it for them. So very quickly items from the game began showing up on the black market.

Real estate was particularly valuable. Describing the size of a virtual world has always been challenging. There is no universal scale applicable to all games. Generally the measure is how much a player can see on the screen. It could be ten meters, ten inches, or ten yards, depending on the size of the monitor and the scale of the pixel. Britannia was approximately one thousand screenfuls; if you took a picture of your screen, you would be able to view the entire world in one thousand pictures. Each screen represented a single piece of land.

At the inception of the game, one screen could be obtained for a pittance; basically, in exchange for game time, you could permanently own a plot of land anywhere in the game. The land had little value because there was so much of it available, but as it got settled, land near the high-traffic areas actually started gaining in value. In the real world you might be able to buy a small piece of land in the New Mexico hill country for $1, but a piece of land that exact same size in New York City would sell for many millions of dollars. The same thing is true in a virtual world, and so the value of property near the entrance to the cities rapidly escalated. A plot of land near the center of town, perfect for a blacksmith shop, might be worth a real $100—because that smithy could make a lot of real dollars selling virtual swords.

The most expensive item was land in a major city; within months of the game going online, a piece of virtual property was sold for more than $10,000 in real money. As the land boom continued, a very smart developer bought a large tract of land in the wilderness that had little value and founded his own city, Pax

Lair, the first virtual city founded entirely by a player in a virtual world. Our game developers helped a little bit, dropping in statues or other elements that a player could not create. While this city did not offer the same protections and services that our cities did, it has prospered and grown since its founding in 1998. It was not established just as an investment, but as an ongoing community.

When we created the game, we offered players many different ways to exist in our virtual world. Players who wanted to have adventures could put on armor and run off to the dungeons or fight monsters and collect gold and bring it back to town, while players who wanted to live a simple fantasy life could fish and hunt, make clothes, run a pub, or forge the armor or weapons for the adventurers.

Blacksmithing was potentially the most lucrative job. But to be successful, your shop had to be near the center of town because that's where the forges were, and where adventurers came to buy new weapons. To operate you needed ore from the dungeons to craft into armor or swords. You had to buy that ore from miners who'd be running back and forth to the dungeon. And unfortunately those miners were being preyed upon by thieves because ore was valuable. The miners needed protection, so they hired fighters to guard them as they transported the ore. The fighters were continually doing battle, so their equipment wore out and the only person who could replace it was the blacksmith. So this nice little symbiotic relationship was created that allowed players—especially blacksmiths—to gain wealth and power. Blacksmiths could earn a ton of gold—but only if people could find their shop. In Britannia, just as in the real world, the three most important aspects of real estate were location, location, and location.

Buying and selling real estate was made possible by eBay, which was just coming into existence and provided a marketplace for whatever players earned in the game. If you earned a lot of gold, you could sell it on eBay for real money, and often, as we discovered, players would take that money and reinvest it in additional real estate. More powerful swords began selling for $5 or $10, and major pieces of armor would go for as much as $100, but attractive real estate could sell for thousands of dollars. It was a futuristic Monopoly, although in this game you could sell the Boardwalk for real money. Eventually several websites were created that reported the exchange rate between *Ultima* gold and U.S. currency. More than a decade after the launch of the game, it remains $7.40 for a million dollars in game pieces gold.

Our game had become an economic engine; it was absolutely fascinating to watch how people played it, changed it, and exploited it. And there was little we could do to stop it. Or frankly, little we wanted to do to stop it. We did have mechanisms in place to stimulate the economy; for example, because equipment eventually wore out, players had to replace it. So there were numerous ways to create value, to generate value, and to take value out of the game. But the economy was always subject to external forces, like an influx of new players, which would inflate prices. A fluctuation in the exchange rates between various currencies as people in different countries made deals also influenced our virtual economy. So a lot was out of our control.

We started hearing amazing stories. Students were paying their way through college with the money they earned playing the game. Criminals realized that this was a good way to launder money—use dollars they earned in their illegal businesses to buy virtual goods, then sell those items online for real dollars. Then we heard about Chinese gold-farming companies. These game sweatshops hired low-paid workers who either played the game for hours on end or wrote scripts for macros instructing numerous AI characters now to hunt for gold, enabling one decent programmer to have endless computer-controlled bots playing the game and collecting everything they could, destroying other players who got in their way. These bots would find the best places to mine value, the quickest and the easiest monsters to kill for the most value. To protect these places, they created very aggressive bots that became very successful at controlling their territories. When real players went into areas where these bots were farming, there was nothing left for them. Initially these bots weren't very intelligent because they were there just to hunt deer or whatever else was necessary to earn gold. So the bots didn't know how to interact with real players. But the Chinese programmers very quickly caught on and created bots that not only were able to defend themselves but would, in fact, attack any "humans" that came into their area. Suddenly these areas of the game map became hyperdangerous for players because of these extraordinarily efficient killing machines.

Other players exploited bugs, finding ways to create duplicate weapons or gold. Glitches and bots "killing" real players worked against our having a fun game, but closing these loopholes wasn't easy.

Like Dr. Frankenstein, we began to wonder what we had created. This real-world economy that had been fun to watch develop was now impeding our ability to keep our player base happy. It was not fun for players to find a mechanical army of bots that they couldn't possibly compete with dominating the most valuable regions of the game. We did make an effort to fight these bots; as soon as a player reported their existence, we could determine what account it came from and close that account. But usually those people would respond by creating another account, often with a stolen credit card. It became very complicated; some of the credit card companies have regulations that allowed them to charge us for unknowingly accepting too many stolen credit cards—even though they had approved them.

Eventually we were forced to ban IP addresses from certain parts of the real world; we banned players from parts of China, for example, because we didn't have enough legitimate players in that part of the world to make up for the gold farmers. The companies responded by setting up relays so they would appear to be playing from another place in the world. We responded by looking at round-the-clock ping times; if they claimed to be in California but it took three seconds for the ping to get there, they weren't in California. It's become a never-ending chase.

We spent a lot of time debating what our proper role was in this virtual economy. Should we support real money trades or try to stop them? Or should we figure out how to monetize them ourselves? This was an opportunity for us to leach more money out of the game. It has been estimated that the economic value of the trading fostered by the games is as much as ten times the subscriber rate. We could have created more game items, like special swords, and put them up for sale for real dollars. We could have created our own exchange, acting as a broker for buyers and sellers and taking a percentage of each deal as our "tax." In fact, there have been games created—especially mobile games—in which players are encouraged to use real money to buy virtual items. Instead of a subscription, they charge money for those items, but rarely do those games allow individuals to trade with other players. Doing so would put them in a very different position, making them a broker, which carries with it a range of legal responsibilities.

We didn't want any part of that. We were selling access to an entertainment service, not magic swords. We weren't a bank, and didn't want to be treated like one. If we were to get involved in that market in any form, even in the smallest way, we would accept a different and far more complicated role and assume additional financial responsibility. Suppose, for example, that you bought a magic sword on eBay for $1,000 and somebody beat you up and stole it from you. That does happen in the game. You paid $1,000 in real money for it and it's been stolen. Who are you mad at? The anonymous person who stole it from you or the company for creating a marketplace that allowed you to lose $1,000? Or let's suppose you paid $1,000 in cash for a sword and there was a server hiccup and we lost a day's worth of play. Servers do go down from time to time. And during that time the purchase of the sword was lost and the seller ended up with both the sword and your $1,000. Who are you going to blame?

Selling objects for real money would have required us to provide a banking level of protection for our players, which we couldn't do. So our official response was that we neither supported nor endorsed this secondary market. We neither recommended nor advised that players spend real money outside the gaming service—and if you did, don't blame us. Players pay us a monthly fee, and for that we promise to leave the servers on for 99 percent of the hours in the month so they have a place to come and play. We make no assurances beyond that.

Since then there have been many other MMORPGs that have been even more successful than *UO.* While those games have been tremendously profitable for the creators, the money made through the resale of virtual items has been far, far greater. And in almost all instances the creators of those games ultimately reached the same conclusion we did. While a few smaller games have done it, the potential legal problems greatly outweigh the financial benefits.

While we were developing *Ultima Online,* we knew that players were going to find creative and very original ways of playing it and that it would evolve, but we never anticipated this aspect of it. This creation of a real economic system completely changed the world of MMORPGs, from the initial conception of a wonderful world for game players to a multibillion-dollar industry.

14 CREATE

We Created a World and Never Got a Day of Rest

When we were creating the world of Britannia, we knew it had to have lakes, rivers, and coastlines. We knew some people would want to live out in the woods, along a river, so we built little homes for them. Logically someone living on the water would have a dinghy, so we made little boats that people could use to paddle out onto the water. And there probably would be a fishing pole on the porch, so we made a fishing pole. According to Richard's Rule, if we put a fishing pole there, it had to do something, so we made fish for players to catch. This wasn't meant to be a key feature of the game, and we didn't spend a lot of time on it, stocking the water with generic-looking fish. We gave fishermen a simple fifty-fifty chance of catching a fish every time they put their line in the water. It didn't matter if they were fishing in a river or in the ocean, there was a fifty-fifty chance. It didn't matter if it was morning, noon, or night, still fifty-fifty. But to our surprise, when we launched the game, fishing immediately became very popular.

Fifty-fifty doesn't mean the result will alternate each time, it means that over a span of time the result will be even. As a consequence players began to speculate that there might be some other rule behind fishing, believing that they had better results when fishing two or three yards farther offshore than when they cast close to the riverbank. Fishing at night, some people were convinced, was more productive than fishing in the afternoon. People created their own mythology; they believed there were special fishing holes, and wouldn't tell anyone else where they were. None of that was true, but players believed it and loved it. Fishing became so popular we started investing money in expanding the fishing simulation; we created different types of fish, rewarded certain techniques, made it more of a challenge, and added more incentives.

The difficulty for us was keeping the lakes and rivers stocked with fish. In fact, when designing Britannia we spent an enormous amount of time, money, and energy trying to create a virtual world that responded in realistically natural ways. Our intent was to keep the ecology of our world in balance. For example, we figured out a way to program the grass to grow: As time passed, the grass would get a little bit taller. Players could actually see the difference if they watched closely enough. That tall grass was part of our carefully designed food chain. It was a consumable for all the herbivores of our world. So rabbits and sheep and deer and other animals would roam around eating that grass, and when they did, it became short grass. There were also carnivores, which lived in the mountains outside the cities but came down to the woods searching for herbivores to eat. They would come down out of the hills and eat the rabbits, sheep, and deer who had eaten the grass. We did a great job of balancing nature: If the grass was too high, the herbivore population grew, which resulted in the carnivore population also growing; but if the herbivores ran out of grass, that population would diminish, and because there would be fewer of them to eat, the carnivore population also would be reduced. And when the carnivores were starving, they would become brazen enough to go into towns and villages searching for food, and sometimes eat villagers.

We wanted players to be aware of this food chain. When the grass was short, there were too few rabbits to keep the dragons sated, so those hungry dragons would attack players. In response the villagers would organize hunting parties to track down the man eaters.

It took us years to create a virtual ecology that forced players to become aware of their "natural" environment and become sensitive to its needs. After a lot of experimentation, we programmed "spawners" to create the precise amount of all the elements needed to maintain this delicate balance. These spawners kept our world populated with whatever was needed, where it was needed, and when it was needed. They would look at spots on the map and if there were no deer, they would make some deer; if there were no lions, they would make more lions, if there were no fish, they would make more fish. We thought it was fantastic. We thought it was an important aspect of our world. Nothing close to it had ever been done before.

Then we invited players to live there. The result was an ecological disaster. A million people ran around killing everything, so right away there were no herbivores or carnivores left to consume the ecology. Players immediately obliterated anything that came into existence. So we had to turn up our spawners just to make sure players had something to hunt.

But no matter how fast the spawners worked, we had problems keeping the world supplied. Players would camp out in the wilderness waiting for game to hunt; no matter how quickly we created animals, they would slaughter them. The spawners were overwhelmed; they were creating one lion at a time, as they had been programmed to do, while ten players were looking for something to kill. There was no way to keep up with the demand. So all this work we'd done to create a virtual ecology was never ever noticed by the players. No one cared about it, no one saw it, no one missed it, so we just ripped it out of the game. All those man years of work we just ripped out and threw away.

We were creating the mechanics of a world, but it was up to the players to operate it. We had some ideas of what they might do, but we were continually surprised. We watched this new society that we had created evolve, and we were fascinated as it mirrored real-life behavior. As I had been trying to do since developing *Ultima IV,* I was creating a game space where the ramifications of your behavior within the game would be reflected to you and enable you to learn more about yourself. *Ultima Online* used a much wider array of methods and allowed players to dig a lot deeper to learn these lessons. It provided them with an almost infinite number of options that forced players to explore their own values and beliefs.

I remember one of the first times we saw players' behavior. This was within the first few months of going online and it was both very surprising and truly gratifying. I played the game as Lord British quite often throughout the first year. I spent the majority of the time trying to help people who were having problems. But I also spent some time officiating at celebratory events like cutting the ribbon to open a new city. Initially some people were still trying to assassinate me, but once they figured out that was futile, most players didn't bother trying anymore. I was immortal, again, so I usually banished those players who tried to kill me to the other side of the planet, and eventually they accepted the fact that it wasn't worth the effort.

Lord British also could be invisible, so I could go anywhere and watch anything without anyone knowing I was there. We referred to that as "gamemastering" because in this mode I could affect the game; I could fix problems, or give people stuff, or take things away, or stop bad guys in their tracks and no one would necessarily know it. One afternoon I was wandering around invisibly, just observing this amazing world we'd created, when I came to a river and saw a character dressed in a straw hat and cut-off shorts sitting on the bank fishing. He was a classic old fisherman just enjoying the day by the mighty river.

But as I watched, an adventurer came onscreen. He was dressed in armor and carrying weapons and clearly had just recently been to the dungeon. As soon as he saw the fisherman, he stopped and said, "Ah, poor fisherman. I see you are without weapons and armor. I have just returned from the dungeon where I have won great treasures. I have more than I need and I will give them to you so that you may start your quest and become more in life than a simple fisherman." Then he began to take items out of his backpack and lay them on the ground next to the fisherman. These were items of significant value: swords, shields, and pieces of armor. They would have made the fisherman much more powerful.

It was an astonishing moment; I was watching a sort of nobility playing out. This was a person who truly understood chivalry. It was impressive—but not nearly as impressive as what happened next.

When the adventurer had finished putting down these items, the fisherman responded, "No, no, you ruffian. What do I want with your implements of war? I am a fisherman. I come out here in the morning and I set my line and catch my fish. In the afternoon I take my fish into town and sell them to the villagers for a modest profit, and with that profit I go to the pub and enjoy my drinks and food with my dear friends and we share the stories of the day. That is the life I lead, that is the life I like, so be gone, warmonger."

With that the adventurer harrumphed and picked up all his stuff and trotted back to the city. It was unexpected and different from any behavior I'd ever seen in a game. I began to appreciate the fact that that this world we had created was beginning to take on its own unique form.

Consider two types of online games. Players in first-person shooters are basically on a battlefield where they run around hunting and killing each other. In many role-playing games though, you're not out to defeat other players, but instead often benefit by working together and forming a team to tackle a monster. Among the things that made *Ultima Online* unique was that, sometimes on purpose and sometimes by accident, there were ways for people to "interfere" with each other. Players weren't restricted to being a good guy; they could choose to be a dirty rotten scoundrel instead. We provided a set of "thief skills," for example, which allowed players to become rogues or pickpockets. Boy, did that skill become popular. We weren't surprised that some players enjoyed being thieves and competed with other players to see who could become the greatest thief. It's estimated that in every role-playing game, whether online or as a solo player, about 5 percent of all players are dissidents or trolls. This means they get their enjoyment not from playing the game in the way it was designed, but from trying to thwart the system, break the rules, cheat, and interfere with other players' progress.

It turned out that while some people loved being pickpockets, other people weren't thrilled about having their pockets picked. Many players simply wanted to enjoy living a fantasy life—to build a beautiful, pleasant, quaint, idyllic, pastoral environment, a place to hang out with friends and speak to each other with thees and thous, and role-play as if they were living in this medieval world. But then a fighter would come in and tear up the place, destroy the furniture, burn down the structure, stomp on the flowers in the garden—and leave graffiti. For the players whose property was destroyed, it wasn't fun at all—in fact it was too much like real life.

Just as in real life, this very quickly became a game of good versus evil, and players could pick either side. It was an opportunity for them to play out their fantasy without suffering the real-life consequences. Some players just waited in the woods for someone else to come along so they could whack them on the head and take their stuff; if they didn't have stuff, they would whack them on the head anyway, just for fun. The dissidents and thieves would devise all kinds of scams and cons, or they would just use brute force. The potential danger added fun as it meant players had to take steps to protect themselves. They carried weapons, used back roads, and traveled by night, avoiding anybody else, or they hired people to protect them until they were safely inside the city with their treasure. The cities were safe, as there were guards all over the town, so when the rules worked as we intended, nobody could steal their stuff. That forced people who didn't want to get whacked on the head to stay in the town. But it didn't stop the bad guys, who realized that since the good guys weren't coming to them, they had to be clever enough to find a way to beat the rules and attack people in the town.

The evildoers were often ingenious in their evildoing. For example, many of them learned the skill of carpentry, which meant they could build tables and chairs and shelving units. What we never anticipated was that those items could also be used as walls. So a group of evil carpenters could each manufacture a chest of drawers at the same time and set them down around another character. There was no way for a character to walk through these chests, so he or she was trapped inside. Then the bad guys would toss a flask of burning oil in the middle, lighting the character on fire. The program-controlled guards didn't notice because they were set to respond to attacks, not to fires. And once the character was dead, the carpenters would move away the chests and take all the dead character's stuff.

Each time we, the creators, figured out a way to make the rules fair and balanced, thieves would find new ways around them. One of the magic items people were able to create was a teleporter, a little blue door that would transport the person someplace else. Players could see teleports, so if they were bringing money back to the city, they wouldn't go willy-nilly into a teleport because they didn't know where they would end up.

But thieves discovered they could hide teleporters just inside the doors of a bank, where it could not be seen from the outside. So the player would safely carry his treasure through the woods and into the city, and just as they believed they had safely arrived at the bank, they would be teleported back into the middle of the woods where a gang of thieves with big axes and clubs would knock them silly and take all their treasure. This cat-and-mouse game in which players took advantage of other players went on for years, and it still does.

So there was a constant battle between people who were trying to figure out ways to take advantage of other people and those other people who just wanted to play the game. The more difficult we made it to attack, the more our group of antisocial players savored the challenge. They weren't specifically interested in stealing stuff, they just got their enjoyment from being creatively evil. But generally speaking, the people who were the prey didn't look at it as fondly. So we had two very different camps we were trying to satisfy.

What happened next was fascinating and something none of us had foreseen. Our players did what mankind has done since prehistoric times—they banded together for protection. They formed groups and associations before going out into the woods. We watched a society grow. We saw communities form. That was one of the magical components of this first generation of massively multiplayer games.

MMORPGs are played entirely in real time. When we first invited players into our brand-new city, the people they interacted with through their avatars were not, generally speaking, people they knew personally. Instead they were people with similar interests. In the real world people live in proximity to each other for many reasons, but not often because they have a shared interest. Going all the way back to tribal societies, people who lived together hunted and gathered together, and they fought together for the protection of the community. In a modern commuting society, that is not the case. Your physical next-door neighbor is not necessarily someone with whom you have any shared interests. However, when people moved into *Ultima Online*, all of their neighbors were people who had like interests. They were people who not only played in the same world, but their virtual jobs and even their virtual hobbies intertwined with each other. New individual relationships formed. People who had never met, who didn't even know each other's real-world names or anything at all about them, became good friends. Cohesive tribal communities were created. There had never been anything like this before.

The strength of these relationships emerged over time. The bonds of friendship and loyalty that were created online proved to be much stronger than anyone could have imagined. Multiplayer games are a shared experience. It's similar to the way in which most people would rather go to the movies with a friend than alone. Let's call that person a member of their tribe, someone with whom they have an ongoing relationship. Generally they don't interact with that person during the film; they sit in their seat and absorb the film. Watching a movie is a passive experience. But having the shared experience of watching a film together provides common ground for a conversation. And an interactive online experience can be even more powerful. Instead of staring at the screen and having the movie play out in front of you, you're involved in trials and tribulations that require coordinated intelligence to succeed and you get to see how different people respond in a variety of situations. You actually get to learn a lot about many different people, and I think that the relationships people build in the virtual world can be very, very deep. People who don't understand that world may scoff at this, but anyone who has been there understands how strong these bonds can become.

There was a group of about a dozen people who played together and all of a sudden they noticed that one of their group had failed to log in. Not just for one day, but day after day after day. This was someone with whom they had shared many adventures. They knew nothing about the actual person—they didn't even know if, in real life, this character was a man or a woman—but they had formed real friendships. And they knew this character was missing. They wondered, and worried, about what had happened to their friend. I'm not sure how they uncovered this mystery—maybe they found someone else online who knew this person's real identity or they contacted our customer service—but eventually they discovered that in the real world, this man had died. Obviously his family hadn't thought about notifying his virtual-world connections. His family might not even have known he had an online existence.

The grief this group felt over the loss of this person was immense. They wanted to celebrate him or her in the game somehow. They lobbied our development staff, asking why the game provided no way of leaving a permanent marker to a person's life in a place that was meaningful to that person and his or her friends. It was an area we hadn't even considered. So our staff got involved and created a custom-made tomb with a beautiful pond in front of it and a dolphin swimming in the pond. Eventually we built tools into the game that allowed players to do it themselves. So now there are graves in the game with the deceased's online name and real name.

A lot of people who haven't played these games too easily dismiss them as "just a game." My answer to that is no, it's not, it's not just a game. The grief these players felt over the loss of a friend was real. It was obvious that the person's human qualities, the characteristics he expressed through his avatar, were so striking that his presence was really missed. We saw that expression of humanity in many different and often unexpected ways.

Clearly players were getting a lot more than entertainment; they were deriving emotional satisfaction from their participation in the game. It's the goal of every programmer to create the most lifelike game possible, to make the gaming experience as close to reality as possible. As I watched that happen, I knew with *Ultima Online* we had come closer to crossing that barrier than anyone before—and by a substantial margin.

Years earlier in SCA—the role-playing Society for Creative Anachronism—gatherings I had seen relationships form between people who played together, but I never expected that to happen in the online gaming world. In the SCA you actually got to talk to people face-to-face, you saw what they looked like, you heard their voices; they were real people wearing costumes. Online gaming was actually the opposite of that; players knew absolutely nothing about the physical appearance of other players. That's one of the reasons we were so surprised when we heard about players getting married.

Avatars did meet and get married inside the game; in fact, I officiated at several online ceremonies. In many instances these were players who were married in the real world and wanted to continue that bond in the virtual world. But what was unusual were those relationships that formed online and then were sustained in real life. When people meet socially in real life, usually the first thing they notice is whether or not the other person is visually attractive. Next they decide if they like that person's personality, and finally they discover the things that they have in common. Online romances are exactly the opposite: By the very fact that they are playing this game, these people have self-selected their relationships. The only people they are going to meet in that environment are men and women who share their interest in playing fantasy games. So they know immediately they have something in common; otherwise they wouldn't be there. Next they get to know the person through the actions of their avatar, and only after a strong online relationship has formed do people even consider finding each other in the real world. The number of people who met online, fell in love without seeing each other, then chose to find each other in the real world and become life partners is stunningly high.

When we originally began promoting *Ultima Online,* I promised it would be "a living breathing magical place, where people could forge true alternative lives." No game had ever come close to actually accomplishing that. Our success in creating a game in which killing monsters or other players or winning a competition wasn't the ultimate goal is summed up by Dr. Castulus Kolo and Timo Baur, researchers who wrote about the game in a study entitled *Living a Virtual Life: Social Dynamics of Online Gaming,* "The dominant motive for playing *Ultima Online* is the social experience of the distributed virtual environment. This is shown in the survey by the fact that about two-thirds of the players mentioned that 'simultaneously interacting with many fellow players' and the 'experience of an emerging society in the game world' are an important aspect or very important aspects of playing *Ultima Online.*"

Especially if they also could kill Lord British.

15 EXPLORE

Finder's Leavers: A-Geocaching We Will Go

I'm an active finder and leaver. Of all the interactive events in which I've been involved, among the most fun and most creative is the international universal treasure hunt known as Geocache.

While *Pokémon Go* has recently taken the world by storm, a virtual treasure hunt has been happening across the globe for years. Geocache was created more than a decade ago to be an interactive experience. Whenever you are reading these words, right now, wherever you are on earth, it is almost inevitable that there is at least one cache hidden within walking distance, and if you're in a city or a town, chances are there are numerous caches within your zip code. There are caches in Afghanistan's Registan Desert and on the summit of Mount Everest. There is a cache in space and at the very bottom of the ocean. The location by zip code of every cache can be found on the website www.geocaching.com, which also will provide good instructions on how to get close to that place and estimate your difficulty in finding it. By the end of 2015, there were more than 10,000,000 people both creating them and searching actively for more than 2,500,000 caches in 180 countries.

Each cache is essentially a treasure chest. A Geocache can literally contain anything; it is something that is hidden and can be found, sometimes easily and sometimes with great difficulty. There is no cost to participate. A cache can be as small as a magnetized film canister stuck to the bottom of a chair in a park or as large as an interactive adventure crisscrossing a city and leading you to a locked warehouse. Items, which have been hidden by other Geocachers, may have value or may be nothing more than a log to be signed. Those items can be kept, left at the original site, or taken and eventually left in another Geocache.

There are also virtual caches, which are simply places to go and check out: "Go to the top of the mountain. It's the most beautiful vista I have ever seen." There are trackable items, which are numbered tags, coins, patches, key rings, or other objects identified by a serial number that can be picked up at one site, registered online, and then left at another, so their movements around the world can be followed. A Geocacher might find a travel bug in New York, then go online and find its history as well as where it is going, then leave it one step closer to its eventual destination, reporting as someone did, "I just found travel bug number 55 in New York's Central Park and I'm going to leave it in St. Lucia." When a Geocacher takes something from a cache, they are encouraged to leave something behind. Each find is registered in a Geocache logbook as well as online.

On the website in addition to the GPS location, there is a description of the item and hints to help find it. For example, a New York cache is described as: "N 40° 45.071 W 073° 58.328 Quiet surprisingly wild area just off the well-beaten path. You may find this place ideal for brief, quiet contemplation. For our very first cache, a small, traditional one, close to home. We place this in honor of our late beloved Teddy, the best guinea pig a little girl could hope to have—and in honor of all the pets we've loved and lost—gone but never forgotten. It is hidden near a busy area, so try not to disclose it. Pleasant Prospect Park location."

A GPS can get a hunter within fifty feet, but then it often requires a diligent search. That's part of the fun. Items have been left in almost every imaginable enclosure. They have been left in tennis balls, in precut logs, in a big red metal elephant on the side of the road, and even in fake dog do. Some of them are so well hidden that even with coordinates, a map, and a description, I haven't found them; I would estimate I've found slightly more than half the caches I've searched for. As a maker of games, I find that extraordinarily frustrating.

My interest in Geocaching began when a woman I was working with named Susan Kath asked me if I was aware that there was an active Geocache right next to my house.

An active what?

Geocache. As she explained the basics of the hunt to me, I was incredulous; something this cool was going on right by my house and I didn't know anything about it? I am a serial enthusiast. When something engages my curiosity or I find something that I like, I embrace it completely. I asked her, "You're telling me that a bunch of people I don't know are coming by my house all the time looking for it?" Well, I instantly set off into the woods. The first Geocache I ever found was a skull hanging from a tree.

A skull hanging from a tree hidden in the woods overlooking my house? You bet I was hooked. The green metal ammunition box at the base of the tree contained a logbook, which described the cache as "Britannia Manor, home of Richard Garriott, the author of *Ultima*."

I was flattered. I began to leave small trinkets representative of objects in my games in the box. And I became actively involved in the infinite search for discovered objects. I have created the lowest and highest Geocaches, leaving travel bugs at the Rainbow Hydrothermal Vents near Portugal and in locker number 218 in the Russian segment of the International Space Station.

By tracking the travel bug I left on the International Space Station, I could see who discovered it and when and who brought it back to earth. It eventually made its way to Geocache headquarters in Seattle, and people are visiting it there at a pretty high rate. Geocaching has actually become quite popular among astronauts.

Thus far nobody has visited the Geocache I made at the thermal vents, at the deepest spot in the ocean.

While the vast majority of caches consist of small and inexpensive items, some are much more involved. One person, for example, built a catapult on top of a mountain that is triggered when a Geocacher sets off a motion detector, which then activates the catapult, causing it to hurl an object over the cliff to land nearby. People have created caches in caves and lakes, in the woods, and of course in the middle of the biggest cities in the world.

Eventually, my property, on which the hanging skull had been hidden, was sold and the new owners erased the Geocache. But by then I had bought other property and had a decision to make: I could wait until someone built a Geocache near my new house similar to the old one or I could jump in and do it myself, I could create an interactive Geocache that provided participants with a higher level of fun. I could do my thing: so I made the search for the Necropolis, as it is known, a day-long interactive hunt across the entire city of Austin. It has been cited as one of the most elaborate and sophisticated caches in existence.

I enlisted several friends, including Dale Flatt, whose imagination has taken me to places I did not know existed. Dale's first career was as an Austin firefighter, but perhaps his greatest skill is being one of the foremost real-world quest designers. He is skilled at creating elaborate, complex, mischievous treasure hunts and mysteries that force people to explore the real world around them. In other words, he was the perfect guy to work with me in creating a Geocache to remember.

Assuming, that is, that you can find it.

With the assistance of several other people we created the search for "The Necropolis of Britannia Manor." It is an adventure that instructs seekers in how to follow a circuitous path through the city of Austin and what to look for: "Be prepared to walk among the dead in their places of rest—be respectful of the dead! Be prepared to enter deep woods and tangled brush, bring with you implements of navigation, perhaps illumination and a camera. Don't forget water, repellent, and other mundane needs of the modern explorer. Perhaps some implements for the defense against the undead might well be advised!"

The Necropolis is a tower about twice the height of a phone booth and four times its footprint. It is always locked. The door on the side has a combination lock. Only Geocachers who have picked up the data along the way on this quest have the combination and can open the creaking door. Inside, there are surprises. It is filled with extremely detailed lifelike decorations, including spiderwebs, skulls, crystal balls, magical objects, and, well, dead things. There is also a quarter crusher that enables people who have made it here to leave with a flattened quarter bearing a memento of the Britannia Manor Geocache. As this Geocache is on one of the highest spots in the area, visitors are encouraged to climb the stairs to enjoy the magnificent view but, as one person wrote in the online log, "I was going to climb to the top of the little tower, but there were tons of spiders on it. Real ones, not fake ones. Just a little too much for me." It has been voted the "Favorite Geocache," the "Best Themed Geocache," and the runner-up as the "Most Mentally Challenging."

Readers of this book who intend to pay their respects to the dead of Britannia Manor should take this code, 618, with them when they visit as thou will need it to open the door to adventure, and to report thy feat! And all who open the door should also try to leave something of themselves to share; among those items that have passed through the Necropolis have been Woody the BiPlane; a St. Michael key chain, "which kept me safe for over thirty years in law enforcement and I hope it will you also while in your possession," and a travel bug, Ying the Panda, whose goal is to be carried by Geocachers through "All forty-eight states, if possible Hawaii and Alaska! Then meet up with his mate Yang the Panda."

I've left seven trackable items. I wanted one of them to go from Austin to Africa and return. It got picked up at Britannia Manor and I followed its progress around the world. Literally, around the world. It made between thirty and forty hops before arriving in Africa, and I don't know how many more before I was alerted to the fact that it was back at Britannia Manor. I went and got it and took it out of play.

One of the first things my wife, Laetitia, and I did after settling into our new home in New York was check to see how many Geocaches were located in our neighborhood. Six sites were found by www.geocaching.com within a five-minute walk and more than a dozen caches within a half mile. Among them were several themed searches that took hunters on educational and entertaining adventures. One of them I found, for example, is part of an elaborately planned city tour that leads people through eight sites that reveal "the political intrigues and armed battles in the city of New York during the Revolutionary War in 1776. Each cache phase consists of a tour of the local area's Revolutionary sites." A lot of these adventure sites require you to answer questions and do calculations to get to the cache.

Hiding in Plain Sight

One of my favorite aspects of Geocaching is that Geocaches are everywhere, literally hiding in plain sight, or at least just under your nose. So I now offer you this little task. Find the 3-D hidden message in this item I found, without attribution, on the Internet. It operates similarly to the 3-D Magic Eye images. Can you read it?

```
lude the as include the as include the as include the as include the as include
ute a right minute a right minute a right minute a right minute a right minute
wn year left down year left down year left down year left down year left down y
mpiler error compiler error compiler error compiler error compiler error compil
econd month i second month i second month i second month i second month i secon
using face her using face her using face her using face her using face her usin
they stop tart they stop tart the stop start the stop start the stop start the
it at wasting pit at wasting pit a wasting spit a wasting spit a wasting spit a
an hast your lean hast your lean has your clean has your clean has your clean h
ead cars time lead cars time lead car time plead car time plead car time plead
 i my go bottom i my go bottom i my go bottom i my go bottom i my go bottom i m
get over frogs get over frogs get over frogs get over frog get hover frog get h
n a pore old and a pore old and a pore old and a pore old an a spore old an a s
rm life rink farm life rink farm life rink farm life rink far life drink far li
server my date server my date server my date server my date server my date serv
back ever fort back ever fort back ever fort back ever fort back ever fort back
 this is filler this is filler this is filler this is filler this is filler thi
up into out of up into out of up into out of up into out of up into out of up i
l never act will never act will never act will never act will never act will ne
lly in the totally in the totally in the totally in the totally in the totally
em a look problem a look problem a look problem a look problem a look problem a
anyone is hide anyone is hide anyone is hide anyone is hide anyone is hide anyo
and distribute and distribute and distribute and distribute and distribute and
```

16

CREATE

Playing for Keeps

One aspect of business I never planned on doing was managing other people. I had enough difficulty managing myself. The earliest game developers largely worked on their own. We created the concept, we wrote the text, we drew each pixel. We were our own engineers, making any beeps or bells or sound samples we wanted in the game. We did our own manufacturing, our own marketing, our own sales pitches. But as the industry grew, we were able to hire individuals and sometimes contract services out wholesale to do a great deal of this work. I'm often asked if I don't pine for the days when one person could create one game that was true to their vision. At some level I do miss those days; who wouldn't want every aspect of a created reality to be exactly as they feel it should be? But it very quickly became obvious that when you're creating vast worlds, a single person can't do it all; they can't even know the whole world! There are parts of every game I've made in the last twenty years that I've never seen or touched and don't even know exist. So sure, I can pine for the days when one person controlled it all, but I get a great deal of joy from being part of a team of people that has been built over the course of years and that knows exactly what I'm looking for but also brings its own ideas.

My greatest skill was programming. That was our foundation, then we did the best we could with everything else. In those early days the way a game looked wasn't even that important; as long as I could trace things out of a magazine or use a simple stick figure, that was sufficient. The earliest computers were not capable of much more.

As computing power grew though, the first truly valuable addition to our team was a talented artist. Then we brought in a musician, someone who could compose music and create sound effects. Again, compared to my primitive beep, beep, beeps, a musician made a valuable contribution. But then it began getting more complicated. Up through *Ultima IV,* with the exception of a few but often critical pieces of code, I did all of the work myself. *Ultima V* began the transition from making games myself to the games being created by ever larger teams with ever larger budgets.

The first substantial specialty group we had to bring in was programmers. Programming was the one thing that the original solo game designers like me knew how to do, but over time the amount of code needed to create a game became far greater than any one person could produce. Programming is an art as much as a skill and, like writers, every programmer has a different style. So when we hired programmers, I would inevitably compare their code, its structure and its results, to how I might have done it. In some cases I would be happy and surprised that someone had done something better than I might have, but as often as not, different meant worse.

Different didn't mean the consumer would see a difference, it might just mean the system operated differently internally so communications between their code and my code would have to be managed in a new way. It was often hard to know if the result would be more or less efficient, but my natural instinct was to force them to do it my way. And when you added a third and a fourth programmer, those problems were exacerbated. I might ask a programmer to make me an AI for a wolf that players could fight. This is a vague request, as there were a lot of different types of wolf encounters; a wolf might run up to you, growl then bite you, or it might follow you for a while before deciding to bite you, or a more sophisticated wolf might follow you for a while without attacking then cry and howl until other wolves arrived and the pack attacked together. Each different solution takes a different amount of time to create, a different percentage of the resources of the machine, and has a different effect on game play.

Each programmer created a wolf in their own way. One of them might spend an enormous amount of time creating an AI system that worked not just for wolves, but also for other animals we might want to introduce later, while another programmer would just make a wolf. When I looked at the result, I had to determine if it would work more or less reliably. There were moments when I became very wistful for the days when I could go into a room by myself and create my own wolf. But that just wasn't possible anymore.

I had to learn how to manage my team, to give clear, completely unambiguous instructions. And once I had defined the goal, I had to learn how to close my eyes and let people do what they had been hired to do. I found that as I became a better manager, the people working for me also became better at their jobs. Eventually it wasn't at all uncommon for someone to deliver something to me that was much better than I would ever have hoped for. They had actually moved the goalposts.

The stakes had become enormous. Suddenly, producing a game could cost hundreds of millions of dollars and take five or more years. If the game was a hit, it didn't matter how much time and money was spent making it, but we had to budget for failure. The art of business, which I had to learn, was how to stay in business long enough to give yourself the best chance to get a big hit.

That meant bringing together all of these skilled people to produce my vision of what the game should be. But if it was going to be my vision, I couldn't expect a director or designer to read my mind. It was up to me to describe things with the level of detail that was required to put all of these people on the same creative page. It never occurred to me that I might not be doing a very good job of communicating what I wanted; in my mind it was clear.

Accepting the fact that I couldn't do everything was really difficult for me to do. It took me quite a while to stop feeling threatened by the fact that there are other really good programmers in this world. I finally became okay with people doing things differently than I would, and not only were their ways good enough, sometimes they were even better than mine. In fact, the skill level required to be a strong programmer today is far greater than the skills I had.

The diverse background I had gained by personally doing every component of the early games meant that few other people have my depth of experience in this business. As the industry has grown and become more complex many people have become experts in their specialties, but they simply haven't been forced to master other aspects of the process. And that breadth of knowledge that I gained probably more through my failures than successes remains essential to designing and leading a team to the creation of a popular product. So, because this is still such a young industry, I often find myself the oldest and by far the most experienced person in any room.

Dollars and Nonsense

That experience had come at a substantial personal cost—even while it made me a fortune. Origin had long been among the top publishers in the industry, but as the industry matured it began consolidating; the bigger players acquired many, many, many smaller companies and became much, much, much bigger. We were always small compared to behemoths like Electronic Arts. I had no complaints about our growth, but EA was so big they had a lot of power over the marketplace. Buyers for the largest game retailers like GameStop, Best Buy, Target, and Walmart didn't have the time to meet with a hundred different game makers to review their games and decide how to fill their limited shelf space. They were interested in speaking only to the industry leaders, the top four or five publishers who controlled a large percentage of the market. Origin was only releasing about four games a year, so if a game wasn't selling, we didn't have an alternate product to put on the shelf to replace it; EA, which released ten games a month, could easily replace it with another game. Maybe we were on that list of top publishers, but only the largest companies were relevant in their ability to control shelf space.

Players loved our games, but sales fell drastically because big retailers didn't have the shelf space to carry them. When you're told by buyers for those stores that they want to carry your games but have no practical way of stocking them because of the way retail business works, you'd better learn how to play the retail game. To survive we had to make a very tough decision. Even today people ask me why in the world we did this and all I can tell them is that in order to compete, we just had to get bigger. We either had to gang up with several other small companies to become a single larger entity or sell Origin to someone else.

We explored both routes. We tried to knit together a consortium of smaller companies, which we then planned to take public. "Taking it public" was always a pretty enticing phrase. It was a complex situation because we were dealing with several companies in different cities and with different agendas. At the same time, we were talking to the companies at the top of the heap, including EA. We'd had some trouble with them in the past, so we faced them with trepidation. Ultimately, it became obvious that we were not going to be able to engineer a merger of small companies and that EA was the right fit for us. We represented an important strategic piece for them because they didn't have any divisions making role-playing games, or that were making leading games for PCs. So in 1992 we sold Origin to EA for $25 million in stock, which we split among all of the top creators in the company. A number of our people became millionaires. It was a bittersweet windfall, but it was the only way we could survive.

We believed merging with EA was the best way to protect the company we had built over the previous decade. The fact that it also turned out well financially was a great side benefit for us. We hoped that this merger would secure our future, and there was every reason to believe we would become a respected and valued piece of EA. They were aggressive and liked to dominate markets, so they wanted us to produce more PC games. They immediately provided the resources we needed to double in size.

This was a huge strategic mistake; we couldn't manage that type of rapid expansion and we wasted a lot of their money. The first thing we did was split our teams in half, which meant that some teams had second-tier leadership. We bought a new building that was twice as large as our existing space and moved into it. Then we doubled the size of our staff. We were rolling! Two years later we realized that about half of those games in development weren't going to be successful, and they told us to abandon them. For us, this was a new and very different way of doing business. Origin had never canceled a single game. Even if we began to doubt the eventual success of a game, we finished it, knowing that at worst the team would have gained the experience of finishing a game and seeing the public response. Going into the next game, that same team would be considerably stronger. But now we were playing by EA's rules.

When we were just Origin, our internal failures were pretty small; we had invested so little money in them and had so few people working on them that when a project failed, we could easily assimilate its staff into another project. That was part of our culture. But at EA there were large teams assigned to each project, so when a game was canceled we had to let a lot of people go. Once, I remember, when we announced that we had to cancel a game in development, the project manager lay down on the floor while another manager outlined his body in masking tape, as if it were a crime scene. Then they put a crown on it and hung a sign, "The King Is Dead. Long Live the King." Over the next few days, people added flowers and candles. We felt this was symbolic of the death of the entrepreneurial company we had built.

Without question the hardest thing for me to do in business was, and still is, to fire somebody. In the early days of Origin, it had rarely been necessary. However, once we became part of EA and grew, then shrank very rapidly, it was much more common. Whenever possible I tried to avoid having to be the guy to do it. What do I know about firing people? I remember my first time very clearly. We had just shut down a project and each of the senior managers had to fire five people. A good friend of mine, a very talented artist, had been working for a project manager we'd fired and there was no room for him on another team. I called him into my office and said something like, "I hate to bring you some really bad news today, but you know we canceled your project and your boss left. We transferred as many people as we could, but I'm really sorry we don't have any more room for you . . ." I couldn't bring myself to simply say it directly, so instead I continued, ". . . and we don't have anything to offer you in the company."

He tilted his head to one side and said, "Well, what are you saying?"

I told him, "I'm saying we don't have a position for you in the company."

"I got that, but what do you mean?"

I kept trying to use subtle innuendo instead of telling him, "Hey, dude, you don't have a job."

He was in such disbelief it took me four or five different attempts before I got the point across. Then he was horrified. He was sad and shocked and angry. He couldn't imagine how I could bring him this bad news. Truthfully, I couldn't figure it out either. Both of us knew how talented he was, so why was he being fired? And why was I the one firing him? He was a friend, so I understood the consequences of this and I was horrified by it. But no one had bothered to teach me how to go about firing an employee.

I learned several lessons from the experience. One, there is no good or easy way to fire somebody. Two, because it is such a nightmare, you have to be much more careful about hiring people, because you don't want to have to go through that experience again. It doesn't matter if it's our fault or their fault, you don't want to bring people into a situation where you might have to let them go. And three, and this was important for me, I realized that if I have to tell somebody bad news of any kind, I need to role-play it first. The few times since then when I have had to fire someone, I've gone into HR and enacted the entire scenario. An HR person would role-play the person I was about to fire, and would try to negotiate with me, make me feel guilty, and become belligerent.

So naturally, having had this experience I accepted it with equanimity when I was "let go" a few years later.

As it turned out, EA was never a comfortable fit for either me or my brother. At heart, Robert is an entrepreneur and had no desire to be one of many minor vice presidents of EA. He wanted to create companies and grow them, so within a year at EA he retired. I just wanted to have the freedom to make great games, so I stayed for a while longer, eight years in fact.

Our greatest success at EA was the creation of *Ultima Online,* which changed the entire industry. In the decade following *UO*'s release, massively multiplayer games—previously a tiny category—generated the greatest percentage of revenue in the industry. Rather than just buying a game once, players also paid a monthly subscription fee, so revenue per player escalated from tens of dollars to hundreds of dollars. Origin gave EA a huge opportunity in this arena, but EA squandered it. We had handed them the keys to the kingdom, but they decided that web-browser games were going to be the future and all but abandoned the MMORPG genre. Eventually they were right—but it would be at least a decade before that happened, and during that time MMORPGs were king.

The reality is that EA earns most of its revenue with terrific games like *Madden Football*. Every year they publish a new edition, which reflects the changes in the NFL. They don't have to create much that's new—they just tweak their football game engine and update the rosters. The rules of football change slowly. At the deadline they wrap it up and release it. The audience is presold.

Conversely, the games we were making could easily take two years or more to create. We released them when they were done. That was not EA's way of doing business. "Richard," they told me, "your release of games is incredibly unreliable." They wanted us to change our development process to meet their deadlines. The game we were developing when we sold Origin was *Ultima VIII*; EA wanted it on the shelves in time for the following Christmas. This was the first time in my life that the realities of business became more important than the quality of the product. They were adamant: "Richard, you need to cut whatever needs to be cut to get this game done." So I cut it; I cut it and I cut it and I cut it, and as a result I shipped the most incomplete, dumb, buggy game I've ever shipped. I still believe that if we had waited until it was complete, *Ultima VIII* would have been a great game. We would have been the first to market it with a variety of features that eventually proved very popular in other games. But we didn't wait, and that was my fault. I bowed to the outside pressure.

Most distressing was seeing the results of making those cuts on both the game as well as my team. The team saw past the warts, knew what we were up against, and loved the game for what it was; they appreciated the innovations in it rather than bemoaning what it could have been. But the press, as well as a number of players, didn't like it at all. The reviews were terrible. All the money I'd been paid had no meaning. I felt awful that I had let down so many people in my effort to be loyal and learn from EA.

And a lot of people had made serious sacrifices to meet EA's schedule. Many of our programmers had worked twelve hours a day, seven days a week for ten months. We would bring dinner in for them because we were afraid if they left, they might not come back. The last month or so we gave them every other Sunday off so, as one of them pointed out, they could see their family or do some laundry. The creative joy we'd once shared in developing a game had been replaced by the prosaic demands of running a business. It was hard to believe how much had changed; only a few years earlier our people would happily work all night and love every minute of it, and now we had become a sweatshop.

At least partially as a consequence of that disappointment, management told me, basically, that they didn't want me making big games like *Ultima* anymore. They wanted very small games and a lot of them. Instead of working on a $10-million game that would take years to make, they wanted games that could be produced in two months for $100,000, allowing them to spread their investment risk.

I didn't agree with that strategy, but hey, EA was my boss. For several years we were able to keep things moving forward, but it strained our relationship even further. Unfortunately, after Robert left, it felt like EA kept sending us second-string replacement managers who saw being farmed out to satellite offices—like ours—as a chance to make their mark. So we would regularly get a new manager who would immediately start by undoing all the things his predecessors had tried to put into place, replacing them with his own initiatives. They would start their own new projects with their own new management tools, fire the junior people who weren't on their team, and bring in other junior people. And within a year these people would get tired of it and quit or be fired. None of them stayed in place long enough to see their new projects through. So every year the entire culture changed. Our inability to manage the expansion, and the succession of B-grade managers, caused Origin to struggle. I was just trying to shield myself from the surrounding chaos and make my next game.

It reached the point where I couldn't even do that. Each time I pitched a new concept to them, they listened politely then told me to go back and change this, change that. Just preparing the pitches could cost as much as $100,000 because the whole process was so bogged down with bureaucracy. I pointed out to them that if they had simply let me go make games, I could have produced five games for that money. But none of my suggestions were accepted. Then one afternoon EA General Manager Jack Heistand called me into his office and said, essentially, "Richard, we just don't need you anymore."

Aha! I understood what he was so inelegantly telling me: "Get your ass out of here, Richard."

I said something like, "Well, obviously you have the right to make this decision. You're the boss and clearly there is nothing I can do about it." All very polite, very rational, two gentlemen conducting unpleasant business. And I left the office, drove to a grocery store parking lot, and wept for several hours.

It was the end of my personal Camelot. This was no game, this was my life. It had been painful for me to fire other people, but as I had just learned, that was nothing compared to being fired myself. I got blindsided by a deep and complex range of feelings. I felt like a failure; I was angry and depressed and confused. I didn't know how to be fired. I didn't know what to do, where to go, how to start rebuilding my life and my reputation. It was a hurt that lasted a long time and, frankly, I don't think I ever fully got over it.

(ATTENTION: Anyone who believes they may someday work for a company in which I am in a decision-making role, please skip the following paragraph.)

When I look at this man who fired me, who, I was told, was himself fired shortly after that, I realize that I should not have meekly accepted this decision. I should have fought for my job. I should have told him to go take a flying leap and started a revolt. I should have told him that if anyone was leaving this company, it would be him. I actually might have been able to pull that off. I should have . . . I should have . . . But in fact I was so shocked at being fired that I didn't.

It was not the last time I would be dumped from my own organization.

(Future employees may resume reading here.)

After I got over the initial pain, I realized this was a wonderful opportunity to work with my brother again. Within twenty-four hours Robert, Starr Long, and I had formed a new company, Destination Games, the idea being that when you leave your origin, you eventually arrive at your destination. I had signed a one-year noncompete clause with EA, so we had time to talk to people in the industry and decide which projects we wanted to pursue. Fortunately for us, during that year EA canceled all of Origin's games, including *Harry Potter Online*. By the time the year had passed, most of Origin's employees had been laid off. This included all the product development people, all of marketing, just about everyone you would need to form a vibrant, creative company. I had worked with many of them for two decades, so it was a no-brainer to say, "Welcome to Destination Games and thank you very much EA for giving back to me for free what I sold to you for twenty-five million a few years ago." Of course, that money was long gone and we didn't have enough to pay everyone, so we suggested that if they committed to working for us, the whole of Destination would suddenly become valuable, and we would find a business partner capable of using all of the services we could offer. So Destination Games opened for business with an experienced, market-tested staff.

Ultima Online had proved there was a huge demand for multiplayer online games so we focused on that market. We were all excited to get back together; we still had dreams to pursue. Within a week of opening the new company, we got a phone call from Jake Song, the Korean programmer responsible for the tremendously popular MMORPG *Lineage*, a medieval fantasy game inspired by *Ultima Online*. His game had been extraordinarily successful in South Korea, counting three million subscribers at its peak, an astonishing 5 percent of the country's entire population. He had been living in L.A. for more than a year trying to figure out how to make *Lineage* work in the American market. Jake told us he was hoping to find other U.S. developers to collaborate with his company NCSoft, and he wanted to find an American distributor for *Lineage*. On paper it was a perfect fit. We were both online game companies. We needed funding, and they had lots of money. They wanted to break into the U.S. market, and we knew it well.

So in 2001, before we began working on our first game, Destination Games was purchased by NCSoft and Jake Song moved to Austin, Texas.

I was extremely excited to be back in my own business, making the type of games I wanted to make. I told a blogger, "So it has gone from retiring and not knowing if you're really going to get back into this industry, or feel inspired to do a game again, to nine months later going, 'I've just got to make more games,' searching to find your way back in, and then suddenly, Bam! Bam! Bam! Bam! Bam! Not only are you in, but you're back on top. I couldn't be happier."

My happiness wouldn't last long. Initially Jake and I discussed creating two separate games, but eventually we decided to work together to try to make one game with both Korean and American sensibilities. It meant that we only had to put together one team, which cut down greatly on our expenses. The game was going to be called *Tabula Rasa*. Jake's game in Korea had outsold *UO* tenfold, so I knew there were things I could learn from him. "I enjoy programming," he said. "My least favorite part is management. I expect the biggest challenge is to harmonize eastern and western culture as I work together with the Austin team."

He was certainly right about that. We tried to work together for about a year. His English was not very good, although much better than our Korean. It was no one's fault, but we had a hard time communicating. We could never tell what he wanted; we'd have design meetings and ask him what he wanted to do and he wouldn't say much of anything. It wasn't that we disagreed or didn't see eye to eye or weren't on the same page; what we had was a failure to communicate! It felt like we were just making stabs in the dark. After about a year Jake just announced he was quitting. We didn't have the slightest idea he was unhappy since we hadn't been able to figure him out. So we had spent millions of dollars on a game that was at least half Jake's that nobody really understood. We were more than a year behind schedule and we couldn't finish the game because we never knew what he wanted it to be.

It was times like this that I remembered sitting alone in an empty classroom doing what I wanted to do—making a game that I wanted to play. If that had been pure creativity, what we were doing now was pure business.

We revamped the game to more closely mimic our original

vision. We backed up and started down a different path, but still planned on it being both a U.S. and a Korean release. The fact was, at that time, as large as the American market was, the Korean market was huge. But this game, while some Asian influences appeared, especially in the art, was shaping up to be more of a western game than a global game. We tried very hard to change that, working closely with the Korean office; in fact several members of our art team were living in Seoul. We kept asking for guidance, wanting to know what elements would help the game sell well in Korea. In response, we continued getting vague feedback from Korea that we weren't getting it, we weren't getting it. What weren't we getting? It. Oh, we weren't getting "it." The Korean NCSoft executives, who had never had a successful game in America, were telling us how to make a successful game in America. I told them, Look, imagine if we were making a European medieval castle game for the Asian market and instead of having nice square blocks as the foundation stones, we had puffy marshmallow shapes. It would look like a comic book version. You want an Asian influence without providing an Asian team. It's not working.

As we learned, the cultural differences were really too extreme for the same role-playing game to do well in both countries. Previously, Origin had successfully sold games in the United States, Europe, and Japan, places where western culture was strongly embraced. But we had little knowledge about Asian cultural influences; in America, for example, the hero of an adventure story is often a tall, square-jawed muscular type, whereas in Asia the protagonist is almost always the meek, mild, nerdy, underpowered character who succeeds despite his physical limitations—and the muscle-bound character is always the bad guy. The small guy's heart and soul allow him to compete, and ultimately defeat the athletic villain. Superman does not do well in Asia.

There were other cultural challenges. Our Asian counterparts would send us pictures from Korean magazines illustrating what the various aspects of the game, from the love interest to the cars, might look like. They went to an enormous amount of trouble to create a series of two-foot by three-foot poster boards with numerous images on them representing character types. One might be titled "Brave women" and have dozens of illustrations of what Asians believed a brave woman looked like. There were posters for "mysterious," "bad boys," "sultry women," "lovers," and other common categories. We carefully placed the boards around the room and looked at them and then looked at each other and realized we didn't have the slightest idea how the representations on one board were different from any other board.

Beyond the obvious differences between men and women, we didn't get the references. Not only didn't we get them, we didn't understand what we weren't getting. That was a problem because they had gone to a lot of effort to make it clear and it made no sense at all. Each side was trying so hard to please the other side, but it just didn't work. We were two years behind schedule and still didn't have a vision that anybody liked. Finally we cut ties with the Korean developers, and set out to create a version of *Tabula Rasa* that would appeal directly to the American market.

This was a mistake as well. We should have closed down that game, accepted our losses, and started over a third time from scratch with a new name, a new delivery schedule, and a new budget. If we had done that, we would have been able to write off the game as a total loss and the new game would already have been partially done with zero dollars spent. Instead, we thought there were some benefits to be derived from keeping the name *Tabula Rasa,* which had some market value in the United States, but doing so meant we were already two years behind schedule and had to carry forward the $20-million loss. People expected a game to come out almost immediately and for little cost, but we knew it was going to take perhaps three more years and many millions of dollars. So we decided to launch the game as soon as we could and continue to grow the game while people were actually playing it. That way we would at least be generating some revenue. It was a formula that had been proved to work. When *Eve Online* launched, for example, it was a quality game but not a serious competitor. But each year they added more depth, more breadth, and more quality, and each year it got bigger and bigger until it finally peaked more than a decade later as one of the best-selling games of all time.

When *Tabula Rasa* came out, we felt we had done the best possible job we could have given the constraints. The game had some brilliant aspects; it was our first serious attempt to tell a story in a massively multiplayer setting. My favorite feature was called "control points," which were regions of the map that often included shops in which players could buy and sell, but also gather information to advance the storyline. But on occasion the game's AI capability would take over a shop, changing the landscape of the game and forcing players to band together to free it so those services would again be available. The AIs were also more mobile, enabling them to move to avoid blows, which changed the dynamics of fighting.

It was published to very mixed reviews, though we were confident we could address the complaints in future updates. But the Koreans were expecting a monumental masterpiece from Jake Song and Richard Garriott, which clearly this was not—yet. Rather than hanging on to it and allowing it to grow while we worked on improvements, the Korean office, which was still heavily invested in it, decided to shut down the entire service.

Truthfully, I very much liked the game and believe if the Koreans had stuck with it, *Tabula Rasa* could have been a strong success for all of us. Even years later I still get e-mails from people telling me that they "got it," and they "loved it," and didn't understand why it was shut down. (Though by all means, it has its detractors as well.)

I had done my best to promote it, even carrying DNA samples for marketing purposes with me into space. But after returning from my flight, while I was still in quarantine, NCSoft's vice president of North American operations, Chris Chung, called and informed me that I was no longer continuing with NCSoft. Astronauts are kept in quarantine when they return from space because their bodies are not capable of doing very much. People stand up and fall down a lot as they readjust. Your brain isn't in complete control of your muscles, so you need a little time to reacclimatize your body to gravity. One thing you are not capable of doing in that situation is business, especially business that involves many millions of dollars.

And then Chris Chung told me that not only was I "leaving the company," they wanted to announce it right away. I asked him to wait three weeks until I could get back to Texas and discuss it with the team. No, I was told, in fact the company had already taken the liberty of writing a public statement for me, claiming that I had left the company to "pursue other interests." Which is what a company often says about a senior executive they let go, but don't want to embarrass.

The letter they wrote, and posted in my name on the *Tabula Rasa* website, hinted that my space flight had opened my eyes to new opportunities. "Many of you probably wonder what my plans are," this forged letter read, "now that I have achieved the lifelong dream of going to space. Well, that unforgettable experience has sparked some new interests that I would like to devote my time and resources to. As such, I am leaving NCSoft to pursue those interests."

A couple of months after this letter was published, the company shut down *Tabula Rasa*—and informed me that they were canceling all my ten-year stock options in ninety days.

When NCSoft acquired my company, they did so for an almost exactly fifty-fifty split between stock and options. I had sold the stock to pay for my flight, so pretty much my entire net worth was in those options. NCSoft cited the public letter they wrote as evidence that I was not fired, but had quit. By "resigning" I was forced to sell those options in a stock market that had been battered by the recession; this was in the winter of 2008, the bottom of the market, and the last thing I wanted to do was sell. At that point my options were worth close to nothing. My contract had stipulated that I had ten years to sell those options, and I wanted to wait until their stock regained some value. I was told that I couldn't because I had quit.

That was outrageous.

I immediately hired legal counsel and sued. In *Garriott v. NCSoft,* we eventually won a $31-million judgment, including interest. Using some of that money, in 2009 we formed Portalarium, with our slogan referencing both Origin and Destination: "We take you there."

All these experiences have made me a much tougher, and hopefully smarter, businessman. My departure from EA and NCSoft taught me a handful of hard-learned lessons, lessons that other people learned earlier and more easily. My most successful games were those games that in my heart of hearts I understood and believed in completely—even if others didn't share my vision. My biggest failures took place when a strong business partner had their own ideas as to what was best for me, and rather than standing up to them, I felt that I could, should, and would follow their lead for this period of time. That has never worked out for me.

I also learned to pay a lot more attention to every word in a contract. That was another new language I had to learn. But now I speak. Every. Single. Word. quite well, thank you.

17 EXPLORE

He's Got a Ticket to Покататься

Growing up in my Houston neighborhood, the concept of space travel was normal. It's what people did when they went to work. I was pretty young for my father's first launch, so truthfully I was probably as excited about throwing rocks at the alligators living in the swamps by the launch pad as by the fact that my father was going into space.

As my father did hold the record for the most time spent in space, we eventually got accustomed to these business trips. While he was on Skylab, our house was wired with something called a squawk box, which basically was a plug-in receiver that let us listen to every conversation between NASA's Johnson Space Center and the crew. It was the raw, unfiltered feed. NASA had warned us that we were going to hear things that the general public would not know about, and that we should be aware that there were going to be systems that failed every day. There is no such thing as a perfect flight, they told us. They didn't want us to be too anxious when we heard someone reporting that the water filtration system had failed or the thruster system had failed, because these were normal problems that they had dealt with before and weren't as dangerous as they might have sounded.

So we would wander around the house listening to the crew

talking to ground control as if it was an ordinary office conversation.

It seemed obvious to me that eventually it would be my turn. I saw no reason why one day I wouldn't go into space just like my dad and his friends. So at age thirteen, when our NASA doctor told me that because of my eyesight I would not qualify to become an astronaut, it felt like I was being put out of the club. This meant I'd never see space. It wasn't like flying to New York; I couldn't book a flight with another company. The idea of private companies flying into space was a Hollywood fiction. In real life only NASA and the Russian Federal Space Agency Roscosmos had ever done it. At first I was crushed but then I wondered who had made this NASA doctor the gatekeeper to outer space? I didn't know how I would do it, but I knew—not believed, not hoped, but knew—that I would figure out a way.

My first real attempt to find my own path into space came around 1987, when I took money I had earned creating games and invested with my father in a company we called EFFORT: Extended Flights for Research and Development. My father had retired from NASA; he had been frustrated with many of the decisions made internally, and saw an opportunity to make things better from the outside. However, as I learned the hard way, just being on the outside does not make it easier. Our concept was quite simple: putting a fuel tank pallet in the back of the shuttle payload bay, which would allow a shuttle to remain in space for more than a month, rather than the limit at that time of two weeks. He believed it would appeal to private companies interested in extending their experiments for longer than was possible then. It turned out that getting it done wasn't so simple. NASA had no interest in it; they were more invested in building a space station than in prolonging existing flights. But while NASA never worked with our tiny start-up, they ended up seeing the merit in what we were doing and built it internally, without us! So EFFORT was not going to get me into space.

Another path appeared when I got a phone call from Buzz Aldrin, the lunar-module pilot on *Apollo 11* and the second man to walk on the moon. He knew that I had invested in companies that were creating the private space industry. Buzz was trying to raise money for his own space venture, a company that would build launch vehicles for private industry. It was a very surreal experience, and an uncomfortable role reversal, to have one of the first people to land on the moon come to meet a twenty-three-year-old to ask for an investment. It had never occurred to me that astronauts were normal people who had to get real jobs and look for funding to build businesses, like any of us! He emphasized that while there were countless people who wanted to fly in space, NASA had absolutely no interest in taking any passengers. NASA was not an entrepreneurial program, it was established to be a space-research program.

For quite some time I believed a company called SPACEHAB would open the door into space for me. SPACEHAB had a plan to build a pressurized container about the size of a double-decker bus that fit inside the cargo bay and was connected directly to the main cabin, allowing astronauts to move freely between their living area and this space. It could be used for experiments—but it could also be used to carry twenty or more passengers. SPACEHAB had an agreement with NASA to build a module. I was the largest individual cash investor at an early stage, and my father and brother also invested. They did eventually build the module and it did fly into space. NASA was very happy to have additional space for experimentation, but never supported the concept of carrying passengers. The last SPACEHAB module flew into space with the *Columbia,* and was lost in that tragedy.

While undoubtedly there were people at NASA or in the defense industry who recognized the need to open space exploration to private companies or individuals, there were just too many hurdles in the way. If there was going to be an initiative to make space commercially viable, it had to come from outside the existing structure. And living in my own pre-Internet world in Houston it didn't even occur to me that there were other like-minded spirits searching for ways to accomplish that dream.

The main lesson I learned from watching experienced space people leave NASA and try to work from the outside, usually without much success, was that even though these people were great test pilots or great scientists and may have had great ideas, they were not necessarily great entrepreneurs. So a new approach was needed.

It was only after I met Eric Anderson, Peter Diamandis, and Mike McDowell that I discovered other people had actively been thinking about this just as I had—and already had a good strategy for solving the same problems.

Our four-man team played historic roles in the development of the private space industry. Peter Diamandis deserves the most credit for his broad vision. Peter is among the most captivating speakers I have ever met; he is an evangelist for an idea, and when he speaks, you just believe it. Space has been his life's work. He understood long before I did that no individual could accomplish this goal, that it had to be a societal undertaking. In college he founded Students for Exploration and Development of Space (SEDS), and then formed the Singularity University—part university, part start-up accelerator bringing together students of diverse ages and backgrounds around the mission to leverage exponential technologies to impact billions of lives. But while Peter's vision was clear, he isn't a closer.

Eric Anderson was the entrepreneur who could build the company and find the clients. He was the person capable of transforming Peter's concept into real dollars. When we needed to find a contact in the Russian space industry, Peter simply flew to

Moscow, walked into the right office, and asked flatly, "What do we need to do to make this happen?" Before meeting Eric the largest investment I'd ever made was in a Golden Corral restaurant or a wildcat oil well deal—which were substantially smaller. I met Eric at the Explorers Club, where he was sitting at a desk handing out cards, telling people, "We're this little company called Space Adventures and we're going to take people into space." I stopped right there. Within a short time Eric was telling me, "Well, of course this is going to work, but in order for us to continue, I'm going to need a hundred thousand dollars from you."

Give one hundred grand to a stranger sitting at a table to do something totally speculative? Excuse me, sir, as nice as your table is, "No freaking way."

Eric couldn't close me in round one, but he knew he had found the right guy. I was thrilled to have found like-minded space enthusiasts who also seemed to display all the great entrepreneurial characteristics you hope to find in partners, so eventually I wrote a check for several hundred thousand dollars—becoming the first large investor and a cofounder in Space Adventures. I pretty much paid for that table and all the little cards. I have subsequently come to understand that when Eric calls, the odds are he is trying to convince me to do something, often part with more money than I am comfortable with to further some scary idea that inevitably will work out.

Mike McDowell is a different beast, but he understood what it was going to take to accomplish the seemingly impossible. He is an explorer who has built several successful exploration businesses. He has climbed Everest multiple times, founded the first company to take people to the North Pole on nuclear ice breakers, and the first company to take people to the South Pole, and then the first and only company capable of taking passengers to deep ocean sites like the wreckage of the *Titanic* and hydrothermal vents. He still runs the only company capable of taking tourists into the interior of the Antarctic. While I have never seen him with a pith helmet on, it's easy to imagine him that way; I can easily see him clawing his way to the top of a mountain, armed with an ice ax and a rifle, valiantly fighting off yeti. As much as anyone I know, he is a person who understands the difficulties of being first.

And then there was me. The key role I actually played in those early days was putting up the money. First, we founded a not-for-profit organization called the X-Prize. To encourage private investment in space, it offered a $10-million prize to the first private company to fly a passenger-capable spaceship into space; at the same time Space Adventures, our for-profit venture, would find and book clients to ride to space. By taking reservations, we would demonstrate that there was a potentially very profitable market for private space travel that would take place on vehicles created through the X-Prize.

I was the first large financial backer of Space Adventures and its largest shareholder, and in its early days I was one of the largest backers of the X-Prize, putting up enough seed money to keep the phones working while we tried to raise the really big bucks.

Space Adventures did take me to the edge of space. I made numerous high-altitude flights in a Russian MiG-25, in addition to several zero-G flights on a Boeing 727-227F Advanced, flights in which I actually experienced several tantalizing seconds of weightlessness. I was getting just the smallest taste of being in space, and during that period it became clear to me that the only way I would ever get there was if we redoubled our efforts to open up space travel to private citizens through Space Adventures.

Providing More Space

One evening in the late 1990s, Eric and I were having a glass of wine at my home in Austin, lamenting how hard it had been for us to raise the $10 million we needed to support the X-Prize, without which no one would create the vehicles we needed for suborbital travel. It occurred to us that we should just skip the suborbital phase and go right into orbit. Eric said, "You know, there are only two organizations on earth that go to space, NASA and Russia's Federal Space Agency, Roscosmos. Let's ask them if they will fly private citizens all the way to orbit."

It was a very nice wine, a wine to dream on. At that moment it seemed like a very logical thought.

Eric approached NASA—which had previously turned us down—and asked politely if there was any way we could buy a seat on the shuttle at any price. The answer was an equally polite thank you very much, but there is no possible way in the world this could ever happen. Their reasoning was quite valid: the American taxpayer, who had invested billions of dollars in the creation and operations of NASA, might object to the fact that their tax dollars were subsidizing a space flight for some rich yahoo. NASA was probably right that this had the potential to be a public relations disaster.

So Eric contacted Roscosmos through our Space Adventures representative in Moscow, Sergey Kostenko, who was coordinating our MiG-25 flight program. We were slightly more optimistic because the Russian government already had a successful "guest cosmonaut" program that allowed foreign governments to pay for one of their citizens to be blasted into space. Since in our mind they were already selling tickets to ride to countries, it was not impossible that they would extend that program to private citizens.

We weren't the only company dreaming about flying people into space. At that time the Russian government, which was in terrible financial difficulty, was also operating the Mir space station. The U.S. government was trying to convince the Russians to de-orbit Mir, which would save Roscosmos a lot of money and allow them to focus their space program on the International Space Station. But a potential competitor of ours, an American company named MirCorp, wanted to buy Mir and convert it into the world's first space hotel. This company actually had paid to send two cosmonauts there to see if it was possible to transform this space laboratory into an orbiting hotel. They had even put up a banner in space reading MirCorp. But apparently the American government convinced the Russians to just de-orbit Mir, not sell it. The Russians had already flirted with the concept of privatizing space and it hadn't worked out at all.

We initially weren't sure about using Russia. Some people questioned the safety of relying on Russian rocketry rather than NASA. Having been raised inside the American space program, I understood their hesitation—but I didn't believe it was warranted. Here's a question: Assuming you wanted to go into space and had the option of a free ticket to fly aboard either the United States space shuttle or the Russian Soyuz, which one would you choose, and why?

Make your choice now: (1) United States space shuttle or (2) Russian Federation Soyuz?

Obviously most Americans would quickly choose the space shuttle, citing superior U.S. technology. Their assumption is that they would be safer on the U.S.-made vehicle. And that assumption would be wrong.

While it is true that American technology is often superior to its Russian equivalent, this does not always hold true when it comes to space travel.

When I make that statement, many Americans instantly disagree with me. But that's mostly pride. There's a reason in my games for not making the obvious choice quickly: Don't let those poor sweet kids out of the cage. Just as there's a reason in nature that things are occasionally very different from what they appear to be: Limestone isn't supposed to be flexible until it is. Ice isn't supposed to form in giant waves until it does. Sometimes things just aren't as obvious as they appear to be.

Russian spaceships are smaller and can't carry nearly as much, and while the crew quarters are very cramped, these rockets are exceptionally reliable. They have been designed for performance rather than comfort. And when I'm putting my life on the line, I'll go with efficient and highly reliable rather than new and improved. Out of necessity the Russians have become very practical. Whenever we've had an accident of any kind, NASA has immediately shut down operations until it can be determined what happened and why to try to ensure that it'll never happen again. They invite independent investigators to do an intense evaluation. And they sort through data, and interview people, and find the problem, and make recommendations as to what needs to be done before resuming operations. At the end there may be new equipment, and a slew of new rules and procedures will go into place to ensure safety. The problem is that they're starting again with an entirely new set of safety regulations, which may not actually have made things safer. While solving one problem, they often create others.

The Russians do things very differently. Their training is extensive—a year spent working in simulators, going through endless scenarios—and during that time you work with the same teacher. We Americans love our technology, but the Russians emphasize institutional knowledge. The history of their space program is one of practicality and incremental changes. If you look at a picture of the vehicle that launched Sputnik half a century ago, it includes the classic core of a Russian rocket with its four strap-on boosters while on top there is a cone and inside that cone is the satellite. Fast-forward a few years to the rocket that blasted Yuri Gagarin into space, and the bottom two-thirds looks identical but they've added an additional booster and on top of that booster a ball containing Gagarin's space capsule. Then go forward a few more years to Soyuz, the vehicle I would ride into space; the bottom two-thirds looks just as it did on the previous two vehicles, and it also has the Gagarin booster with the escape and habitation module on top of that. So while it has been slowly improved since the first vehicle that ever took anything into space, the lineage is very clear.

In contrast, the United States's first vehicle was a Redstone ICBM booster that was not capable of going into orbit. We were behind in the space race, so we rushed to stick Alan Shepard on the tip of an ICBM and shot him downrange. Once we got past a couple of *Mercury* flights, we switched to *Gemini,* a whole new rocket system built from the ground up to take people into orbit. Once we had people in orbit, we decided we wanted to go to the moon, so we scrapped *Gemini* and built the entirely new *Apollo* system. After traveling to the moon, we went back and built the space shuttle, which was radically different from anything we'd done before. While of course there was some carry-forward into the next generation of American space vehicles, in many ways we were starting from scratch. The problem is, when you only fly an experimental vehicle a handful of times, you're not able to shake out the bugs. And so if you then start over at zero again with a new vehicle system, you have this enhanced profound risk that comes back with each restart, which they never did in Russia.

The Russians couldn't afford to start over, so they made incremental changes as necessary. I couldn't have fit into the original Soyuz capsule. But through the years they've made it more efficient, and they've made it bigger.

The Russians are continually making small improvements. The Soyuz that I would be launched in had a modern glass cockpit like any state-of-the-art airplane. It included digital computers with all the bells and whistles. Right beside my knee in the craft was an oxygen control valve. It has two handles that can be moved up or down to add oxygen into the airflow to keep the crew alive. If it's ever needed, it's a pretty important piece of equipment. And when I went back and looked at the very first Soyuz, the T1, the exact same oxygen control is in the exact same place, made by the exact same manufacturer. In theory the Russians probably could have saved half a kilogram by reducing the size of the handles, since they don't have to be that big to do the job, but it was never done. That valve had never failed and they didn't have new technology that would prompt a change. The function hadn't changed, it had never failed, and there hadn't been any changes in the requirements. So they just left it alone.

The International Space Station (ISS) reflects the way our two cultures approach similar projects. The ISS is a grand global partnership in which as many as fifteen countries have built modules that are then sent up to the space station and attached. The ISS consists of twenty or so school-bus-size pieces that have been built over almost two decades; none of them ever mated to another on the ground. About the only thing they have in common is that they were all made to match a standard docking port. In fact, no one really knows if a module is going to fit until it gets there; they cross their fingers and send it up. Fortunately, thus far all the pieces have fit together.

Moving through the space station from one section to the other, you see the evolution of technology since the launch of the first module in 1998. Starting from the stern, the first and second modules and the air locks leading from them were built in Russia. Next is the first American segment, followed by a mix of U.S. and other partner-built modules. The U.S. segments are at least six feet across; they are so wide that I can't touch both sides at the same time, while the Russian segments are about four feet wide. But the differences run deeper than that. The familiar silhouette of the space station includes giant solar panels, as big as a football field, that extend out on giant booms. All that is U.S. power; the Russian segment has far smaller panels that are barely noticeable.

Different modules are used for different functions. The Russian end of the space station contains all those systems that keep it alive and functioning. All the food lockers and heating capability are in a Russian segment; it is where the galley and all the food preparation equipment is located, all the freshwater is stored and the wastewater is purified. The only space toilets are in a Russian segment, and trash is shot out into space from a Russian module. The two original sleeping quarters, the air lock used for spacewalks, the docking mechanism for supply ships, the cargo storage area for equipment not in use, and the batteries in which the energy from the solar panels is collected are all in Russian segments. The Russian modules are the industrial living quarters and they look that way; cables are strewn across the hallway and all the works necessary to keep the station alive are in there.

Then you float through a hatch and you're in the American segment. There is no confusing the two. The American segment is bigger, brighter, and pristinely white. It's antiseptically clean. Because experiments are being run there, no one is permitted to bring food or drink into that side. The experiment racks are big and well-lit and carefully separated and labeled. To leave the U.S. and U.S. partner segments you travel through a zigzag tunnel that gets smaller and smaller; the walls of this tunnel are covered with water storage bags, so it feels like you're inside your intestines, and the lights get dimmer and the equipment is sort of yellowing with age, and finally you emerge inside the Russian cargo bloc. This is where people brush their teeth and damp-towel their hair, so water has sprayed on the wall and over time become moldy; it's dingy and creepy and has the feeling of being in a haunted house. It's like being Cinderella in space, floating from the grand ballroom back into the hut in the forest.

The Americans on the ground are always pointing out the superiority of the American equipment, while discounting the fact that they are completely relying on Russian equipment to stay alive and get the work done. Maybe their equipment isn't as bright and shiny, but it works, it has always worked, and there is no reason to believe it will suddenly stop working.

But when we initially approached the Russians about Roscosmos, their response was *nyet*. Basically, they explained that it would simply take too much time to determine if it would be possible: how would they train civilians and how much would it cost? We already knew that other countries were paying about $20 million in cash or trade for a seat, so we figured the price would be in that range. We asked them how much we'd have to pay them to figure out if it was possible. The answer was $300,000, and we knew they could take our money, do the study, and then determine that it wasn't feasible.

Red Rider

Space Adventures had essentially no money to back that study. But as I intended to become the first private citizen in history to fly in space, I provided the financing to cover this $300,000 fee. This was my dream—it was what I wanted to use my wealth to accomplish. Not that I'm stubborn, but I was going to show that NASA doctor who'd said I couldn't fly in space. Three months later Roscosmos informed us the cost for an individual to be launched into space to live on the International Space Station for two weeks was exactly what we had guessed, $20 million. While my purpose in going into space was to show that private spaceflight can be commercially viable, to show that there are unique business opportunities that ultimately will drive that demand and eventually create a commercial industry, truthfully I also was fulfilling my lifelong dream.

I immediately booked my flight. At long last, I had pulled it off. I had earned enough money to go, and helped build the company that managed to buy seats on a rocket: I was all set to be the first civilian to travel into space! Take that, NASA doctor!

At the same time this was happening, I was building my dream house in Austin. It was an intricately designed giant fun house, complete with secret passages and hidden rooms. It was very expensive to build and seemed to get more expensive every day. But I owned a lot of EA stock and it continued to go up, just outpacing the cost overruns on my house.

And then, right before I was to begin the formal training process in Russia, the Internet stock market bubble burst, taking with it most of my money. In a few weeks my net worth decreased from tens of millions of dollars in the bank to one of millions. This was the most profound loss of my life. I had spent thirty years getting to this point; I had helped found the X-Prize, I had helped found Space Adventures, I had arranged with Roscosmos to fly, I had the money to pay for the flight—I had my ticket into space. If the dotcom bubble had lasted one more month, I would have sold my stock and paid for my seat.

But it didn't. The crash was completely unexpected. Who knew this bubble was about to burst? There have been times in my life when I wept, there have been times when I curled up protectively, but I didn't know how to deal with this. I couldn't pay for the flight or finish the house. I was wiped out. I was devastated, crushed, absolutely totally completely crushed. I had to abandon the two biggest dreams of my life. Wow, Richard, I thought, while wallowing in what was still a nice amount of money so long as you didn't intend to make a trip into space or build a wildly expensive house, this has all been for naught.

Having reached an agreement with Roscosmos, Eric and I did not want to lose this opportunity to further our business, so as disappointed as I was personally we began searching for someone to buy my seat on the Soyuz and become the first private citizen to fly in space. Besides, I assumed that somehow I'd eventually figure out a way to pull this trip off for myself.

We approached a former scientist from the Jet Propulsion Laboratory, Dennis Tito, who had gone on to become one of the founding fathers of quantitative investing and had formed a a large and successful hedge fund. Dennis had been involved in the MirCorp venture and clearly was as passionate about traveling into space as I was. I didn't know him at the time, but he agreed to buy my seat and so made history.

At first, I dealt with this incredibly painful situation by ignoring it as much as possible. I blocked it out of my mind as much as I could and didn't even watch the launch. I knew he was going, but I stayed busy. I guess the good news was that I was already so disheartened, his success could not make me any sadder. I don't believe Dennis Tito knew the whole history of his flight nor the fact that in my mind he was sitting in my seat. For a long time afterward, in fact, I thought of him as my nemesis, the guy who stole my seat without having to do any of the work to arrange it. It was only several years later when we had an opportunity to spend time together that I discovered he had been pursuing this goal for at least as long as I had, and frankly had spent more time and even more money trying to create the MirCorp flight opportunity before me. So if there was anyone who had earned that first position, it was Dennis Tito, who subsequently has become a friend I greatly admire.

Even if he did take my seat.

Apparently NASA had tried to convince Roscosmos not to go ahead with this, and when that failed they sent an astronaut to Russia to try to convince Dennis Tito that his flight would be bad for America. It was a bizarre accusation—nothing about it would be bad for the United States—and the flight went ahead as scheduled.

While the flight was personally devastating to me, it was a huge success for Space Adventures. We had successfully opened the era of private space travel. Following Dennis Tito's flight we launched four more people into space. By that point we had founded Destination Games, which was a reconfigured Origin, and two weeks after forming the company had sold it to NCSoft for a substantial sum. My net worth was rebuilt and once again I could afford to buy a seat—although the price had increased by more than 50 percent.

Finally, it was my turn. After making a large initial payment, I had my first physical. The medical review boards of both NASA and Roscosmos have to approve anyone flying into space. I went through the initial medical test without any difficulties. I have some minor medical anomalies—for example, both of my kidneys are on the left side of my body—but there appeared to be nothing that would prevent me from flying. Incredibly, someone with this same condition had already flown without any difficulty, although his condition was discovered post-flight. After making a second payment, I had to go through considerably more testing. This may well be the most extensive physical examination in the world; the structure of every artery and vein in my body was probed. And that's when doctors discovered I had what is called a hemangioma on my liver. This meant that one of the six lobes on my liver did not drain properly; it was a genetic defect of some sort. On earth this wasn't an especially serious problem, and it hadn't caused me any issues, but in certain situations in space, such as a rapid depressurization, it could have caused fatal internal bleeding.

A flight medical specialist called me and told me he had good and bad news. You actually can learn a lot about a person by discovering whether he or she wants to hear the good news or the bad news first. What about you? If a doctor called and told you he had good and bad news, what would you want to hear first? The bad news, he told me, was that this hemangioma was a medical issue that would disqualify me from flying. Once again, I was devastated. I had been so close so many times.

By this time I had many millions of nonrefundable dollars invested in the flight. There is no thirty-day return policy at Roscosmos. And there was no longer any medical insurance for this. One of our earliest passengers had been able to buy insurance to cover just this sort of eventuality. After he had spent several million dollars, it was discovered that he had some scarring on the inside of his lung, caused many years earlier by its collapse. As a result he flunked his flight physical, meaning he lost his reservation slot and millions of dollars. He was able to recover all the money he'd spent, and as a result no company would underwrite another policy. So if I couldn't fly for medical reasons, the many millions I had paid were gone.

That passenger had a surgical procedure to remove the scars and about a year later successfully flew into space. The good news for me was that my hemangioma could also be surgically corrected—the really bad news was that the operation was life-threatening and not recommended. We found a doctor who specialized in the operation, and who cautioned me, "My advice to you would be to not do it. Stay here on earth and live the rest of your life happily. But if you are determined to do this, I guess I can justify it by considering it a professional requirement."

I didn't ask anyone for advice and I wasn't interested in hearing it. There was no option as far as I was concerned. Dennis Tito could not have this operation in my place. I never had the slightest apprehension about the surgery and it never occurred to me that it wouldn't be entirely successful. The surgeon operated as quickly as it could be scheduled, as the recovery period was six months and I had to be in Russia to begin training.

When I awoke in Houston's MD Anderson Cancer Clinic, the doctor was standing by my bed. "Welcome back, Richard," he said. "The surgery went fine. And by the way, while we were in there I noticed your gall bladder wasn't functioning properly, so I removed it."

What? You did what? I was absolutely horrified. We had never discussed my gall bladder and I had no idea if that would disqualify me from going into space for reasons I couldn't imagine. Does an astronaut have to have a gall bladder? What does a gall bladder do in space? Fortunately, my flight-medicine doctor reassured me that a functioning gall bladder was not required for my journey.

The Die Is Cast

In space, there is no room for a backseat passenger who just sits there. In a very cramped space capsule, every inch has to contribute and my body occupied a lot of inches. So I had to be trained to become part of the crew, and that required moving to Russia. I actually spoke a smattering of Russian, a remnant of my parents having spent time there when my father was working for NASA. So I knew enough to find a library and a bathroom. While I would be okay if we were flying to a bathroom in a library, I would not have the slightest idea what I was being told if I was warned in Russian, whatever you do, don't touch that lever! And that would have potentially meant flight casualties. So I began studying Russian, and while I never became proficient enough to read Tolstoy in the original language or carry on a conversation, I did learn enough to read and understand my flight manuals as well as any instructions that the Russian flight team would throw my way.

I trained in Russia with American astronauts, who immediately and graciously embraced me and never made me feel like an interloper. We trained on the vehicles we would go up and down on as well as all the modules that comprised the space station. As much of the training as possible was done in Russian, because all of the instruments are in Russian as are the mission control commands. The instructors are all bilingual, but as the course proceeds it becomes more Russian intensive.

Training required an entire year. During that whole time I experienced only one brief moment of apprehension, and that took place at the space launch facility in Baikonur. We had trained for months in a launch vehicle simulator but this was the first time we were going to sit inside the actual capsule. It wasn't mounted on a rocket, it was sitting on a test stand. The simulators always had a big hole in the side, which made entering and exiting very convenient, but entering a real Soyuz capsule was far more complicated. We had to go through a side hatch into a habitation module, then shimmy down through the main hatch to the center of the descent module and finally into our seats. The first time I did this was on a very cold October day. There was no power running to the vehicle, so no lights were on inside; it was absolutely black. I was wearing my space suit, which not only prevented me from looking down, it was so thick it made it almost impossible to feel things. After sliding down the hatch, I had to rest my feet on the edges of the commander's seat, then use my feet to feel my way over to my seat, while at the same time being extremely careful not to kick the seat belts because if they fall behind the seat, it is impossible to find them. Instead you have to climb out of the vehicle to let someone go in and reposition them, and then start the entire entry process again. And this is just to get into your seat.

I was proceeding very slowly, very cautiously, practically scraping my nose on the hatch until I was finally able to scoot down into my seat. It was dark and cold and incredibly cramped. There was equipment all around me and I could barely move. I remember thinking that if I were at all claustrophobic, this would be an absolutely terrifying experience. My second thought was the realization that if there was an emergency, if the capsule caught fire or we landed off target or there was a depressurization or any other unexpected event, there was no way I was getting out. Entering and exiting under ideal conditions was difficult; doing it when there was a problem was not going to be possible. So I accepted the fact that if there was an emergency, I was going to die.

There wasn't anything I could do about it at that point. I was already physically, emotionally, and financially committed to this flight. I was very philosophical about it, deciding, Well, I certainly hope things go well because if they don't, this definitely is going to suck.

I did know things could go very wrong. As a child I had lived through a potential space disaster that few people outside NASA knew about. The news media never reported how close my father's first flight came to disaster. During the era of the *Apollo* program, things often fell off vehicles as they were launched through the atmosphere. On an *Apollo* space capsule, there were four clusters of four-thrusters, a total of sixteen, one cluster on each side of the capsule, which were used to position and control it. By firing some combination of these thrusters you oriented the spacecraft; it enabled the crew to rotate it or pitch it in any direction, or to move forward or backward or from side to side. But to have complete command authority of altitude and position in space, you had to have all four clusters working.

During my father's launch two of those four lost all their fuel. They were 100 percent disabled, with no chance of recovery. It's considerably more difficult to control the spacecraft with only two clusters of thrusters; it's very hard to line up with the space station and dock. It requires great skill, but theoretically it's possible to pull it off. It also complicates reentry and requires the crew to sort of back into the atmosphere without burning up, but again, it could be done. The larger problem was that a third thruster on my father's capsule was losing fuel. Losing that third thruster makes it impossible to orient the capsule for reentry into the earth's atmosphere. If the crew could dock prior to it leaking out all the fuel, NASA would have the chance to send up a rescue mission. They immediately began speeding up the third Skylab flight vehicle for a potential emergency rescue mission. I don't remember any sense of panic or fear; everybody just did their jobs professionally. NASA was very matter of fact about it: if they have trouble with that one, we'll spin up this other one to pick them up. There was never any sense of Oh my God! We've got to get them now!

Even though we heard every word as we walked around our house, we were confident that they would solve this problem because they always did. In fact, they managed to shut down the fuel leak in the third cluster. Not only did those guys manage to dock that thing, they didn't lose any additional fuel and carried out their mission nominally.

So I understood how easily things could go wrong, but I also was confident that they could be corrected.

Another No Go

I have tremendous respect for the men and women who created the space age—but there are parts of the bureaucracy that have evolved with it that are extremely frustrating. By the time I got ready to fly, five private citizens had already flown to the ISS, and I had been part of the NASA family for most of my life. During my training, I had spent two weeks at NASA to ensure that if I ventured into the American segment of the space station I wouldn't break anything, and to learn the emergency procedures in those modules. I had even agreed to conduct several experiments for NASA while living on the ISS. So I was more than a little surprised a few days before my scheduled launch when NASA made a final effort to stop me.

It has never been clear to me why the agency created to open space for mankind would actively try to prevent access to it. But this was far more serious than a missing gall bladder. Before being launched into space, everyone has to sign a Space Act Agreement. This is a sort of contract that basically says we know you're going to be up in space near our stuff and we'll let you do that in return for doing things for us. My Space Act Agreement committed me to doing some work for NASA, including compiling data samples as part of several long-term experiments it was conducting. In return NASA would support my work with the Challenger Center, an educational program established to encourage young people to get involved in space science. That was really important to me. I had been a founding member of the Challenger Center and planned to very publicly support it during my flight. About three months before the launch, I signed it and went on my merry way to continue my training.

Two weeks prior to the flight, I was put in medical quarantine in a hotel in Kazakhstan. Three days before the launch, a NASA representative, astronaut Mike Baker, suddenly showed up and informed me that NASA hadn't cosigned the agreement, and moreover, someone had noticed that it didn't quite include all the clauses they suddenly believed should have been included in it. It was nothing, I was told, just read it and respond and everything'll be fine.

I explained as politely as possible that we were taking off in seventy-two hours and they were requiring me to sign a legal document without access to legal representation. Didn't it occur to anyone that I was in quarantine in Kazakhstan?

Four new clauses had been added. First, rather than performing the experiment I had volunteered to do for them in the American module, which is how I had trained for it to be conducted, they now wanted it done in the Russian segment. That seemed pretty silly, as it would adversely affect the quality of the data I was collecting for them. Second, I would not be allowed to take any photographs through the American windows of the ISS. While that made no sense to me, as there are about fifty windows throughout the space station, it was not a problem. Perhaps they were worried about me scratching their windows. Third, I could not take any photographs of the interior of the U.S. portion of the ISS. That was completely ludicrous. There was no logical reason for that; there had already been countless photographs and videos taken of every foot of the ISS. Space Adventures' first five flyers had been permitted to take pictures, and this would make it very difficult for me to document my own trip. I was furious about this one—but only until I read the last clause, which prohibited me from entering the U.S. segment of the International Space Station for any reason whatsoever.

It was the ultimate no-trespassing statement. It was a clause they'd sneaked in at the last possible moment and it made the trip as difficult as they could make it. There was no way I could sign this document. I went to officials and to my crewmates to ask for guidance. I was advised not to sign it, just to get in the rocket and go. But Mike Baker was persistent. Mike wasn't the bad guy at all, and I'm quite certain this wasn't his doing, but he had been charged with carrying the message. I made excuses as long as possible, until he warned me that if I didn't sign the contract, NASA was going to do everything possible to prevent me from making this flight, and if I did fly, they would withdraw their cooperation with the Challenger Center and reject any of the experimental data I compiled.

"Let me make sure I understand you," I said. "You're going to throw away data worth millions of dollars that I'm providing for you at my own cost if I don't sign this?"

Yep.

I was furious. Incensed. I spoke with a crewmate who calmed me down and suggested I sign it—but only after adding a clause of my own that reiterated the first rule of the space station: the commander of the International Space Station has the right to modify any rules he or she sees fit while in orbit. The captain of the ship gives the orders. I cited this clause in the agreement then signed it.

Two days later I finally made it into space. After the Soyuz had docked with the ISS, American astronaut Mike Fincke, who was becoming the ISS commander, said to me, "Richard, welcome to the U.S. segment of the International Space Station. Do your work where originally planned throughout the ISS; please don't ask me again for permission. Have a nice day."

And indeed I did.

[Of course, for the NASA lawyers who may be reading this, Mike Fincke did originally consider the possibility and ramifications of restricting me to the Russian segment of the ISS. However, the two sleeping quarters there were already spoken for, and the remaining area either had no climate control or was a high-traffic spot at all times of the workday for both the U.S. and Russian crews, and so, therefore, it was decided that my work and experiments should be conducted in the originally planned and agreed-upon areas of the ISS. To do otherwise would have compromised efficiency and would therefore add both cost and risk to U.S. astronaut activities. I am sure you will find that in his official report.]

18

EXPLORE

Packing (and Smuggling) for Space: The Space Kielbasa

With the airlines encouraging people to pack all their belongings into small suitcases, the question becomes: What would you pack in preparation for two weeks in space? Room is severely limited, so travelers on the Soyuz are restricted to about five kilograms of up-and-down weight. My standard kit consisted of three pairs of underwear, two T-shirts, a pair of shorts, and my flight suit, as well as toothpaste, toothbrush, and a comb. In addition I was permitted to bring what is essentially a small shoe box of personal items. That included everything I needed to conduct my experiments and any other personal items. I had to make my selections very carefully; in effect, each kilogram cost about $40,000.

As I was to learn, there are restrictions. I met Mike Fincke for the first time in a little Italian restaurant named Frenchie's. It's an astronaut hangout right next to NASA, but I had never been there. There are signed pictures of all the astronauts who have been in space hanging on the wall, including one of my father.

My father was with me for this meeting. We were enjoying a nice wine when I casually asked Mike, "Does anybody ever take any alcohol up to the ISS?" Before he could answer, my father said, Spockishly, "Richard! What an outrageous question. Of course not, that's completely against the rules. It would be totally irresponsible. It would be dangerous and interfere with the oxygen scrubbers and a lot of other systems."

There was a long list of items besides alcohol that I was not permitted to bring. For example, I wanted to bring an iPad, an iPod, an iPhone, or even a Microsoft watch tied to an Outlook calendar, just some type of PDA smaller than a laptop so that I could listen to music while I worked as well as have a reliable calendar to help me stay on schedule. None of those items were allowed, I was told, because they hadn't been certified—and it costs several hundred thousand dollars to get any item certified. NASA and Roscosmos were concerned that a battery might overheat and catch fire, but it was also pointed out to me that these items have a glass face and if that glass broke in space, it could create shrapnel—which at zero G could be very dangerous. Okay, no alcohol, no batteries, no glass. Got it.

While my father was on Skylab, he had filmed several experiments with magnets showing that if you let a magnet free-float it lines up with the earth's magnetic field. Above the Northern Hemisphere the north end points down, above the Southern Hemisphere the south end points down, and over the equator the magnet floats horizontally. I thought it would be both interesting and fun to repeat those same experiments. But I wasn't allowed any magnets because they could possibly erase the hard drives of the ISS. I thought this was ridiculous, since most people have magnets in their homes and offices and I've never heard of anyone accidentally erasing a hard drive. But that was the rule. Okay, no alcohol, no glass, no batteries, no magnets. Got it.

Many people had suggested experiments I might conduct during the trip. A high school student asked me to place a laser pointer at one end of the ISS and aim it directly at a sheet of graph paper at the other end. That way, when there was movement on the space station, someone jogging on a treadmill, for example, it would be possible to determine how much flexion there was in the length of the ISS. I was kind of curious about that myself. So I asked permission to take a pen laser with me. Lasers were not allowed, I was told, because I could burn out somebody's eye or mess up other experiments. Okay, no alcohol, no glass, no batteries, no magnets, no lasers.

But eventually "anonymous" members of my crew gave me some good advice. "Richard," I was told, "you don't need to take any of those things up there with you—because they are already there." And that's when I began to learn the glorious history of "smuggling" items into space. Long before I'd thought about it, astronauts had already taken magnets and lasers—and other supposed contraband—into space with them.

Most often this was done with the assistance of a member of the crew support staff. There are quite a large number of people involved in the process and all of them have precisely the same objectives: make certain each flight is as safe and productive as humanly possible. And a distant third, try to help everyone have the most memorable and successful flight possible. It was common knowledge that if a member of that support staff wants to work with you, right before you board the rocket there is a handoff moment when he or she can hand you something to put in your pocket. Or, more commonly, before you zip up they can put an item inside your suit.

However, in a previous flight that handoff had caused a potentially serious problem. Every astronaut has a pressurized spacesuit made specifically for them. Long before it is actually used in a flight it is tested and retested numerous times. But during preparations for the launch of an earlier flight, as a cosmonaut's spacesuit was being pressurized, a zipper broke, causing the rubber interior layer to bulge out into the capsule like a big balloon.

It didn't burst, though, and it survived to the required pressure. Mission control had to make a serious decision: should they scrub the launch, meaning they would probably have to make a whole new spacesuit, requiring them to cycle back that flight two or three months at a cost of tens of millions of dollars, or should they risk launching with a damaged spacesuit that would be needed only in the unlikely event that there was a serious malfunction. They decided to go ahead and fly.

Officially no reason was ever given for this malfunction, but several months later I was given an explanation that—as farfetched as it seems—is likely too farfetched not to be the actual cause of why the zipper had busted. During the handoff a member of the capsule support staff had secreted a kielbasa between the outer and inner layers of the cosmonaut's suit, which was then zipped up. It was like sneaking candy into a movie theater. But during the preflight pressure test, the smuggled kielbasa exerted significant additional pressure directly against the zipper, causing it to break. Since "the kielbasa incident" it has become far more difficult to secrete those few extra tidbits in your equipment.

Coincidentally at the same time I made this flight I had been working on the new game, *Tabala Rasa,* which was loosely based on a traditional futuristic concept that the planet had been destroyed and only a few survivors had escaped into space. In that game these survivors were responsible for restocking humankind, which obviously would be created in their own images. Suppose, I imagined, that actually happened; which people would we want to be our ambassadors to the future? If I took their DNA with me and civilization was destroyed, the ISS would become a Noah's ark for humanity. Sometime in the future an advanced civilization could use this information to reconstitute humanity. So we stored digitalized DNA samples from selected people on a device known as the Immortality Drive. The Immortality Drive would also be a repository of the knowledge and wisdom and achievements of humanity, to be left on the space station just in case humanity got wiped out. It was a great idea; it enabled us to promote our new game as well as possibly save humanity!

The big question was, who to invite to be a part of it? Stephen Colbert came up very early in our discussions. Not only was I a huge fan, I had seen that he was very interested in all things space. On his show he had already created a variety of fake medical products, one of them touting something called Formula 401, which is basically his semen. The pitch was that if women were in need of a donor, who better could they possibly get? If he was bold enough to suggest that his DNA should be used in that way, it made sense that he would also believe that if someone's DNA had to represent humanity in the future, surely his would be a very good fit.

He was more than willing to participate. We sent him a DNA kit and he returned a swab of cells taken from the inside of his mouth. It was sequenced in a genetic lab, then digitalized and encoded on our Immortality Drive, where it was designated Formula 4001. "I am thrilled to have my DNA shot into space," Colbert said, "as this brings me one step closer to my lifelong dream of being the baby at the end of *2001*."

In addition to Colbert we stored the digitalized DNA of other noted humans, including Stephen Hawking, screenwriters, musicians, Silicon Valley icons, and Matt "The Blueprint" Morgan, the seven-foot-tall 320-pound wrestler on *American Gladiators*. He earned his nickname because he was considered "the paradigm of human genetics and physical potential." I transported the Immortality Drive to the ISS and left it there when I returned to earth, where it serves as a possible foundation for the future of mankind.

It could happen.

I also "launched" Commander Montgomery Scott's DNA into eternity. I had received a phone call from a friend in the States telling me that Chris Doohan, actor James Doohan's son, had tried three times to fulfill his father's wish to have his DNA sent into space but had failed. James Doohan had created the legendary *Star Trek* engineer, Scotty. I am a big *Star Trek* fan and I'd met James Doohan at several conventions; I'd found him to be a genuinely likable guy whose work I appreciated. There is a for-profit company that will carry ashes into space and Chris had tried three times and each time their rocket had failed to reach orbit. My friend told me it would mean a lot to the family to see James successfully launched into space. I also thought it was a wonderfully appropriate gesture.

A good man named Dale Flatt has been my property manager in Austin for more than three decades. Dale's uncle, Paul Forbis, had died at a very young age and to honor him Dale had done a very cool thing. He'd printed business cards with his uncle's name and the dates of his birth and death on them, then put just a bit of his uncle's ashes on each card. Whenever a friend of Dale's traveled, he'd ask him to carry one of these cards with him. I'd taken cards to Antarctica and to the *Titanic,* and I intended to take one into space. Using Dale's idea, I made three laminated business cards with a smudge of James Doohan's DNA on them and put them in a Ziploc bag with several other cards people had asked me to carry into space.

I took three cards because I had a plan for each one. Number one is still on the space station, so Scotty will be circling the earth indefinitely. The second card I used when I began my trip back to earth. When you're descending out of orbit, one of the first things you do is jettison the module in which you lived for two days on the way up. There's an air lock between that habitation module and the descent module; a screen on the habitation module prevents anything from blowing back out toward the descent module when you undock. With my crew's permission I put a card between a hatch and that screen, knowing that when we jettisoned the module, that card would be flipped into space. I put the second card in there, and when we detached the habitation module I sent James Doohan on an eternal space walk. The third card I held on to and gave to Chris.

There were some experiments I was not able to complete. A project suggested by a student that seemed interesting was to put a slow-motion camera in front of an ice cube on a stick and film the ice cube as it melted. Instead of the melting water dropping onto the floor it would turn into a sphere of water. To do that I would have to put some water in a freezer and make one ice cube on a stick, but NASA refused to allow me to do that. Those freezers are American assets and they would have charged me to use them.

Sometimes NASA's restrictions were counterproductive. When my father was on Skylab almost thirty-five years earlier, he had conducted a photographic survey of the earth. I intended to take those same photographs so we could see how the earth had changed in more than three decades: how the rain forest may have been degraded; how pollution may have increased; how proper forest management techniques may have brought some areas back to a healthy state. We put a program together in cooperation with the Crew Earth Observations team at NASA. Volunteer scientists from the Nature Conservancy helped us look through the Skylab archives to identify possible photo targets. In doing this I helped develop a piece of software that has become a standard tool used by astronauts to identify places on earth.

Before my flight, officials from the Crew Earth Observation team sat down with me and admitted they were concerned that the chips on which my pictures would be stored were going to be off-loaded by the Russians. They thought it would be faster and more reliable if I just dropped them in folders and e-mailed them directly to NASA from space. I could send them down on the TETRA satellite networks, they said; TETRA is a gigantic data pipeline that had been built in anticipation of a shuttle program that never happened. It had tremendous bandwidth that was barely utilized.

Even though it would take up a considerable amount of my time in space, I agreed to do it. If I could contribute to science, it would be worth it. Then they told me, "Oh, Richard, because you're a private citizen, we're not allowed to just give you the bandwidth." Apparently Congress had mandated that any private citizen using it had to pay. Well, that's fine, I agreed, but just how much does that bandwidth cost?

Without blinking they told me the first bytes of data would cost me about $350,000, and then it would be more per megabyte. The same organization that had refused to let me make an ice cube in their freezer also wanted me to pay a substantial fee to provide research data for them. I'm sorry, I told them, there's no way I can justify that.

I brought my photo chips back to the ground where they were processed normally by the Russians, then returned to me. I handed them over to NASA and, by the way, all the pictures came through just fine.

While most of the restrictions were imposed by NASA, there was one thing I wanted to do aboard the ISS that the Russian Roscosmos had some problems with—I wanted to create a piece of art. I knew that if I just drew a picture in space it would basically be a pretty bad picture painted in space because I'm not much of an artist. So I began trying to figure out how to make art in space that is remotely interesting. Eventually I came up with an idea I thought was brilliant: I would cover the interior walls of a container with paper, then release drops of paint and let them migrate through the container until they settled on that paper. I was playing with watercolor paper that actually is slightly hydrophobic, which means it doesn't absorb paint. So, in theory at least, the paint would touch the surface of the paper and sit there. Because there is no pressure, those drops of paint wouldn't flatten out. If I left it there for a week or so it would dry out like a raisin. I was going to make dots of colored paint on paper, but my belief was that the profile of those dots could not be scientifically replicated on earth. It truly would be space art.

But as I was learning, nothing about traveling into space was simple. The Russian Federal Space Agency is very methodical. They want everything done in an orderly fashion. When I sat down with agency officials and told them I intended to release colored liquids on the space station, the initial response was why? Why do you want to do that? This is a problem, I was told. We cannot let people paint the interior of the space station.

But the Russians also are logical, fair, and reasonable. Rather than rejecting the concept they worked with me to figure out how to do it. It turned out to be considerably more complex than I ever would have imagined. The Russians actually helped manufacture the various pieces I needed. We ended up constructing a plastic container about the size of a glove box that could be folded

Making art in space.

flat. We made small triangular plastic bags and filled them with tempera paint. I could cut off a tip with a pair of scissors to release the paint. Got it.

But when I cut open the packets of paint and waited, nothing happened. Without gravity the paint didn't flow or drop. If I squeezed the packet, it made a jet of paint or created a gob of paint on the tip of the plastic. I had to gently shake it off, which meant I was imparting velocity to it. As a result these paintings all have a discernible splatter pattern, because that was the only way I could get the paint off the packet.

19

EXPLORE

Taking Up Space

On October 12, 2008, I found myself sitting in one of the three seats on a Russian Soyuz TMA-13 rocket preparing to fulfill my lifelong dream. It had cost many millions of dollars and years of my life to be sitting in that seat at that moment, and as I sat there, listening carefully, I heard absolutely nothing.

There are a number of traditions every astronaut respects on the day he is launched into space. I began by signing the door of the room in which I would stay. Then I was blessed by a Russian Orthodox priest. I put on my launch suit and joined my crew for the ride out to the launch pad, during which we watched a movie entitled *White Sands in the Desert*. It has absolutely nothing to do with space flight, but has been a part of Russian space history dating back to Yuri Gagarin.

The road to the launch site is in terrible shape, full of potholes; the Russians like to joke that in honor of the great Gagarin the road he traveled to his rocket has been left in the same condition. During the drive, as the bus passes behind a hill that blocks it from the reviewing stand, the male passengers get out and relieve themselves. I don't know if this tradition also dates back to Gagarin. That was also the place where small items have been known to change hands—until the unfortunate incident of the kielbasa.

At a NASA launch site, no one other than the astronauts and the small group of people who help load you in and seal the rocket up behind you is permitted within five miles of the rocket after it is fueled. The Russians are a little more flexible. We were greeted at the launch site by a band, in addition to several Russian dignitaries—and my father. As he escorted me to the rocket, a heavy mist of liquid fuel and kerosene was rolling down its sides. The whole situation was surreal.

As I took my first steps up the gantry, a Russian general kicked me in the butt, another carefully followed ritual. I stopped with my crew and waved good-bye, then got into a tiny elevator. We took this rickety elevator to the loading area and climbed into the rocket.

I was the first one in; I squeezed into my seat and powered up the vehicle as my crewmates joined me. We began going through out checklist. There is a scheduled one-hour pause-in-place before the launch. During training we skipped that because it wasn't necessary; there was no point in sitting there doing nothing. So we had an hour with absolutely nothing to do. We chatted pleasantly for about fifteen minutes, then ran out of things to say. So we sat there in silence, and then someone in mission control turned on elevator music. It was almost impossible in those long minutes not to think about everything that could go wrong.

Then we went through the final commands. I turned to the last page of my manual and at the very bottom is the Russian word for *start*. As we worked down the page toward that word, I was surprisingly calm. I had lived this moment a thousand times in my mind. I was content.

It suddenly occurred to me that liftoff actually happens seconds after the engines throttle up. We lit the engines. Then we kept going to the next command, and the next. I was most surprised by the immense, overwhelming silence. I thought I would hear the engines roaring into life, but I didn't; instead we were enveloped in an almost eerie silence.

Then we hit the button, and my dreams came true.

I also had anticipated feeling a jolt as we lifted off. But as the rocket began to rise, I felt only a tiny vibration. Gradually, I began to be aware of the push of gravity against my chest. I felt pressure pushing me back into my seat, and it continued to increase. I was stunned by how smoothly we made the transition from static to motion. It was less like a car taking off at a stoplight than a ballet dancer being gently lifted into the arc lights.

There was little time to appreciate the beauty of this feat of engineering during the eight and a half minutes before we reached orbit and the engines cut off. There was a lot going on inside the craft as we scurried up to four and a half Gs of pressure, and went through two staging events. But somewhere deep inside I was feeling incredible joy: I had done it. I was flying in space.

There were two cameras in the cockpit and our checklist noted precisely when each of us would be seen on a camera. Even before I flew I knew that this would be a moment to be used to send a message. The back pages of my flight manual faced the camera, so anything written there would be visible. I wrote a message in the symbolic language of my game *Tabula Rasa*. And in this pictographic language I quoted from the Russian scientist considered the father of rocketry, Konstantin Tsiolkovsky: "Earth is the cradle of humanity, but one cannot live in the cradle forever." And when I knew the camera was focused on me, I showed that page.

Our engines cut out at the precise second the G-forces disappeared, and rather than being pushed into my seat, I was completely unrestrained by gravity, held in my seat only by belts. I would not feel the constraints of gravity for another two weeks.

As our vehicle began rotating, I looked out my small window

and got my first view of the earth from space. While I might have been in awe of the beauty, or thrilled by the fact that I had accomplished my goal, in fact my first thought was, Wow, we're not really that high up. I had the sudden realization that we had better be in a perfectly circular orbit or we would be going home a lot sooner and a lot warmer than expected.

My journey had begun.

Life in Space

Life on earth does not prepare you for living in space. Many basic human functions have to be done differently in space. Breathing, for example, or sleeping. In zero gravity you don't have to lie down on a bed when you go to sleep; in fact, you can't lie down because you'd float in the air currents. Astronauts can sleep anywhere on the space station; they crawl into a sleeping bag tethered to the floor, a wall, or maybe the ceiling.

Even breathing inside a spacecraft can be very dangerous if you're not moving. Before you continue reading take a few seconds and think about that: Breathing is the single most natural function we perform. Nobody has to be taught how to breathe on earth. Babies do it instinctively. So why is it different—and potentially life-threatening—in space? And how would you solve this problem?

When you take a breath on earth, you exhale carbon dioxide, which is warmer than the air around you so it tends to float up and away, allowing you to breathe in mostly fresh oxygen. In space there is no convection current that naturally moves expended CO_2 away from you. Instead, because there is no "up," that potentially dangerous gas just hangs there in front of you. So when you're not moving, you're breathing in that same pocket of air. And when you're sleeping, you're not moving. Scientists solved this problem by installing huge fans that continuously blow air through the entire space station, to be filtered at the other end. But sleeping and breathing are simple compared to going to the bathroom.

Instructions for Using a Toilet on Earth

Preparation:
a. Open toilet lid
b. Stand or sit down (optional)

Use:
a. Deposit liquids or solids directly into the bowl

Shut down:
a. Flush

Going in Space

"How do you go to the bathroom in space?" is the most common question kids ask astronauts, but it's also the most common question new astronauts ask experienced astronauts. There is actually even a book on the subject, titled *How to Go to the Bathroom in Space*. But I can now assure you that there is nothing in that book, nothing learned in training, and nothing told to you by experienced astronauts that properly prepares you for the complex reality of going to the bathroom in space.

Long before entering orbit you are taught the theory behind every system and get to practice almost all of them. To learn how to operate the radios, for example, you sit down with those radios and make calls on the actual flight hardware exactly as it is done aboard the space station. The only thing you don't practice by operating the actual hardware is the toilet. Generally, this is because it is the only piece of equipment in which gravity plays such an essential role in its operation that you can't duplicate it on earth. On earth, gravity pulls human waste down into the toilet. In space, with no gravity, there is no up or down, and we are not capable of simulating this operation on the ground. So instead, future astronauts examine the hardware and learn how it is supposed to be operated in space. Unfortunately, all the discussion is about the sanitized version.

The ISS space toilet was invented by the Russians. In many ways it is a marvel of engineering. In fact, when the original space toilet had to be replaced, NASA sent up a second Russian toilet because none of the toilets invented in the United States are superior. This space toilet works, although how it works is not nearly as simple as I thought it would be.

I was taught this theoretical method for using the space toilet. A space toilet looks nothing like a normal toilet, even an upscale normal toilet. The toilet on the space station is about the size of a short telephone booth. In this telephone booth there are two control panels. An aluminum drum that looks remarkably like a beer keg and serves as the receptacle for solid waste (that's number two, he wrote sanitarily) is bolted to the floor. There is a box about the size of a shoe box with a lid on it mounted on top of it. When you lift the lid, you see the only piece of wood that exists on the space station, the toilet seat. There is a hole in that seat into which you are supposed to put a plastic bag. The bag is there to collect solid waste. It goes down into a soda-can-size hole. The bottom of this bag is perforated. Coming out of a wall next to this space-age toilet is a vacuum hose, which is connected to the bottom of the beer keg and creates suction that draws the bag from the shoe box into the keg.

Liquid waste is easy to dispose of; you just take the tube off the wall and pee into it. There are two funnel attachments, a round one for men and a lozenge-shaped one for women. You just hold it an inch or two in front of your body and urinate into the tube. The air draws it through the wall into a centrifuge that separates the liquids from the gases. The liquids are recycled into the drinking water and the gases are filtered into the cabin air. No, you can't taste it. Functionally it works without difficulty, and it's so easy to learn even an astronaut can do it!

Solid waste is a completely different issue, and potentially a dangerous one. On the space station this bathroom is right next to the galley. That's important for two reasons: First, when people aren't working, they are usually hanging out in the galley, so everybody sees you going in and, eventually, coming out. And second, you're warned that the most dangerous system on the ISS is the toilet because if any feces were to escape into the station, it could contaminate the drinking water, the food, and everything else. During the class the instructor really impresses upon everyone how important it is to do this correctly and, literally, to not mess up.

In fact, during training in Russia when they told us what to do and not do, they also related a space-toilet horror story. It was a big deal: On a previous flight an unnamed astronaut had come close to contaminating the entire galley and the entire crew had to stop doing whatever they were working on and join the cleanup effort. So I listened to the instructions carefully because I did not want to screw up and become part of future lesson plans.

Here is how the system is designed to function: When you go into this bathroom, there is an array of switches you have to turn on, which operate the machinery that treats all the waste, from pretreatment all the way through disposal. Once inside you situate yourself over the cola-size hole and do what comes naturally. And do it into the plastic bag. Your waste is drawn into the bag by the air suction flowing into the keg. When you're done you remove the rubber band holding the plastic bag in place and the bag is drawn into the container, then you put a new bag in place. After the container is filled, the bags of poop are sealed into the keg and the keg is ejected into space. Eventually it reenters the atmosphere and can be seen burning up by people on earth who would describe it as a shooting star.

Remember that the next time you look up into the sky on a beautiful night and are lucky enough to see what you think is a shooting star.

That's the way the system is designed to work. That's the way you think it's going to work. That is not the way it works.

The practical application is not so easy. First, it's necessary to understand what happens physiologically as your body processes solid waste. On earth as you move around, your body is actually helping your gastrointestinal tract process food, taking out the nutrients and leaving the waste product. It happens on some kind of roughly predictable time schedule. That time schedule in space is very different. Because you're floating rather than walking or climbing or running, your metabolism slows down as much as fivefold. So, hypothetically, if you go to the bathroom once a day on earth, in space you would need to go to the bathroom about once every five days.

But in space, after five days, when you do need to go, you really need to go.

When describing what happens, rather than using the word *poop* or sanitized phrases like *number two,* I prefer to use toothpaste as a metaphor. The imagined visual is much more appealing and little kids understand it much more easily. And so do adults. On earth if I held a tube of toothpaste over a sink horizontally and squeezed the tube, toothpaste would flow out and gravity would take hold and eventually rip it away from the tube and drop it into the sink. But if you were to do that same experiment in space, toothpaste would come out of the tube and stretch a half inch, an inch, two inches, four inches, as long as you kept squeezing it

Instructions for Using a Toilet in Space

12 October 2008 Richard Garriott Page 19

АСУ МКС – Toilet (СОЖ 4.1.2)

Prepare:

ППС23	↑	Toilet power on
ПУ1	↑	Panel Power on
	✓↓	Manual off
	✓↓	Pre-Treat 2 doses
		Show Status
	⊙□	Observe Urine Collection System lamp
ПУ2	↑	Auto
	✓↓	Manual

Liquids:

Funnel		Remove cover & Open Stopcock
ПУ1	✓□	Sep Normal & □■ Pretreat Dose
	○	Check Suction
		Utilize, Clean, Discard Wipe, Cover, Stow

Solids:

ПР		Open Cover
	✓	Check for Insert
		Use, Clean All, Discard Insert & Replace
		Close Cover

Shut Down:

Funnel		Close, Cover and Stow
	⊙□■	Observe Separation Normal <= 23 seconds
ПУ1	⊙■	all lights off
ПУ1	↓	Panel Power off

Changing Urine Tank:

	Open Panel 138
	Disconnect hose at P3 of ЕДВ-У
	Loosen 4 lower bolts w/ driver,
	Remove 4 upper bolts w/ wrench
	Turn 4 brackets 90deg to release housing
	Remove cap P3 of new ЕДВ-У, recap old, discard
	Connect, Install, secure, mate, Close 138
ПУ1	Manual
	Electric Reset No: Sep/UR Tank Full
	Manual off

Changing Solids Tank:

Loosen 2 wing nuts, Remove, close cover, tighten
Remove hose, close cap
Loosen 2 base bolts, remove and discard container
Reverse with new

would continue to come straight out because there is nothing to draw it down. It's going to float horizontally. In fact, if you were to move the toothpaste tube, the line of toothpaste would move with you because toothpaste is really sticky. It holds together very well.

The first time anyone uses the space toilet, they do it with some trepidation, but also with confidence. You're an astronaut, living on arguably the most sophisticated piece of machinery ever devised. Based on the experiences of every person who has flown in space, brilliant engineers have worked for decades to create a toilet system. So you go into this bathroom and power it up, carefully following the instructions. It's been days since you've gone to the bathroom, so you're ready, believe me you're ready. You situate yourself over the shoe box and within seconds you've created a column of "toothpaste" that remains connected to your bottom but also reaches the bottom of the plastic bag. And then, as you keep going, you begin to feel resistance. This is the part of the curriculum not taught in bathroom class. So you do what feels natural, you stop.

And then you start thinking, Okay, what do I do now? If you stand up you're going to drag this tail with you into the bathroom and you have been warned many times that the one thing you absolutely cannot do is let anything get loose in this bathroom. So standing up is out. Second, eventually you have to get up. Third, you can't keep going because that will cause a bigger mess on your backside without resolving the situation. So you sit there considering all the possibilities. You know there has to be an answer because it's impossible that you are the first person to encounter this problem.

I began looking around the room, wondering if there were any tools that might be useful. There were only two movable objects in this room other than me and the toilet: little plastic pouches containing ordinary wet wipes and rubber gloves. Wet wipes and rubber gloves. Think about it.

The first thing I did was what, I later discovered, is known to all astronauts as the "bounce maneuver," an action in which you bounce up and down briefly on the seat. The downward inertia causes the part that is attached to you to break free and end up in the plastic bag. At that moment I experienced a feeling of great joy, believing I had solved this problem.

In fact, I had not, not at all. A memorable image from the classic motion picture *Caddyshack* is that of a Baby Ruth candy bar floating in a swimming pool. In fact, you actually don't have to have seen the picture to be able to visualize the emotional impact of that scene. So now imagine that candy bar diagonally across the plastic bag, which means that the plastic bag is no longer empty and available for further deposits. It now has an obstruction in it.

It does occur to you to wonder why, during the long months of training and tests and teaching, that no one had ever mentioned this situation to you.

Since the bag has an obstruction, it will take considerably less volume to reach the bottom, and after several days you have no shortage of volume, or perhaps "toothpaste" inside you. So eventually you end up discovering the secret of the space toilet, a maneuver that almost everybody finally figures out. You put on a rubber glove, take some of those wet wipes, and manually push the obstruction out of the way.

And then you repeat the process as often as necessary until you finally finish your business and clean yourself up, throw the rubber gloves in the bag, take the rubber band off the bag and seal it and drop it into the keg, then set up the system for the next user. By the time you're ready to leave, forty-five minutes of slow, careful, manual manipulation has passed and you're probably sweating because of all the work you've had to do. But as you leave that room, there is one brief moment of great pride as you realize the obstacles you have managed to overcome, and that lasts until you see all your crewmates gathered in the galley laughing;

and after another few seconds you start laughing too, knowing that you have just completed one of the most difficult obstacles of the entire journey into space. You have completed an important initiation ritual and have joined an extremely select group who have used a toilet in space.

There are many lessons from the development of the space age that are important to remember as we move into the digital future. Sometimes just because we are capable of doing something doesn't necessarily mean we should do it. New and bright is not necessarily better than old and dim.

That, and my satisfaction that I didn't screw up too badly going to the toilet in space and thus won't see my name—as others did—in the hall of shame Russian manual that reminds all new space flyers which of their peers they don't want to follow the lead of. I had not shamed myself, or America.

The Sounds of Silence

One question I am often asked about life in space, even before the success of the movies *Gravity* and *The Martian,* is what is the real danger of objects hitting the ISS? Officially the answer is, are you kidding? It's really dangerous out there. At least two to four times each year a piece of space debris large enough to be life-threatening comes close enough to the space station that the ISS has to be moved out of position to avoid it. But there are also tons of smaller space junk that can be tracked by radar and that are continually hitting the ISS. Space junk doesn't have to be especially large to be dangerous; both the object and the ISS would be traveling at about 17,000 miles per hour when they intersect. That's about 25,000 miles per hour relative velocity, and energy is the square of velocity, so the square of that velocity times the mass of an object equals a tremendous amount of energy. For example, a piece of a bolt about the size of a pencil eraser would pass right through an astronaut on a spacewalk, killing him instantly. Think of it this way: a bullet travels at just over mach 1, and the space station travels through space at mach 25, so a bullet fired in space would have twenty-five times the power that it does on earth.

One simple way to determine how often the ISS is hit by debris is to look out its windows. The ISS consists of a series of connected modules supplied by different countries. A month before I arrived, for example, the Japanese Experimental Module was added to the space station. Because it was new its windows were big and perfectly clear. There was not a scratch on them. But in the Russian module, which has been in space for more than a decade, the windows are pockmarked. It's been getting hit for more than ten years. Those dings are made by objects no larger than a paint flake and they average about one hit a year on each window.

On my second day in space, I had an experience with subatomic particles being blown around by solar winds. I'd heard about this happening. These particles are atomic nuclei, consisting essentially of one proton and one neutron, the heaviest parts of the atom. On earth we're protected from these particles by the magnetic field and the upper atmosphere. In space there is no protection from these invisible particles, so they pass right through your body, usually without any interaction. Occasionally though, there is an interaction that actually becomes visible. Scientists don't know precisely how this happens. One night I was almost asleep, and my eyes were closed, when I "saw" one dot suddenly get brighter and brighter and brighter and then dimmer and dimmer and dimmer until it faded away into darkness. It was an incredible experience and woke me up. It was as if a flash had gone off in my brain. Almost instantly it went from overbright to very dim. I didn't know what it was and obviously I was very curious. I checked with astronaut Mike Fincke and he'd had the same experience. We had been hit by radiation that was so strong we could see it.

So there is a lot of microjunk flying around in space and the ISS does get hit by it. Another night I was sleeping in the European Columbus Module, which is amazingly quiet, as opposed to the Russian end in which several very noisy survival systems are in continuous operation. As I was about to fall asleep, I heard an unusual sound. It was like nothing I had ever heard before and nothing I had been briefed to expect. It is very unsettling to be in outer space and hear a sound you aren't supposed to hear.

It was a soft sound, vaguely like heavy, wet snow hitting a car's windshield. It was a wisp of a sound and it didn't last long: about twenty minutes. Whatever that was, I thought, it's gone. Then it came back again. I heard it again the following night, and every night after that, always at different times and only for those few minutes. Every night I'd try to trace it to its source before it disappeared: Was it coming from an air-conditioning vent? Was it made by an instrument? Was it made by the expansion and contraction of the module as it went from the sunny to the shaded side of the earth? It intrigued me, and infuriated me because I couldn't figure it out. In the silence of the night, I would float around the cabin trying to identify it, placing my ear as close as possible to every potential source. It seemed to be coming from the bow side of the module's skin. Once again I asked Mike Fincke if he'd ever heard it and he had, but he had no idea what it was either. I've speculated that it could be tiny space debris striking the bow, but that seems unlikely, as that type of debris would quickly sandblast holes in the ISS, so I just don't know. Whenever I have the opportunity I question other astronauts about it. It remains a mystery to me, but I haven't given up the possibility that one day I will find that answer.

20 EXPLORE

Don't Kill Stephen Hawking!

I'm not generally known for freely expressing my emotions. I guess I got that from my father, our family Spock. My father has the emotional range of a scientist, which essentially can be described as how high he can raise his eyebrows. He always has been a logical, rational man who chooses each word very carefully. When we were children if we were asked a question, for example, we couldn't respond by saying, "Right," because right is also a direction that makes the response ambiguous: did we mean the direction or correct? He insisted in that situation that we use the correct word, *correct,* if that's what we meant.

So that obviously made an impression on me. In my casual speech I'll still use that scientific filter and in most emotional situations I'll remain quite calm. I remember once a colleague pointed out that I didn't seem to be as bothered by some bad news as everyone else in the company, saying, "You never really seem like you just want to go out and have a beer," he said, "and just be sad."

But even I became emotional when I got to watch Stephen Hawking on a Zero Gravity flight. The whole thing wasn't my idea; Peter Diamandis, my partner and the cofounder of the Zero Gravity Corporation, which sells the ZERO-G Experience—

flights aboard a modified Boeing 727—thought it would be an amazing event to let the world's expert on gravity escape gravity. Beyond the obvious medical issues, the problem we faced was that this flight would be very expensive. My main contribution to this effort was to join several other people on the board of directors to contribute the money necessary to make it happen; in return we all got to participate on the actual flight. The board thought it would be great marketing—that was our way of justifying the expense—but the truth is, we all thought it would just be a very, very cool thing to do.

As most people are aware, cosmologist and theoretical physicist Stephen Hawking suffers from a motor neuron disease that gradually has left him completely immobile; when I met him his only means of communication was by moving a cheek muscle as a computer cursor slowly moved over word and command options to indicate where it should stop. The muscle in his cheek is almost the only one in his body he can voluntarily control. When he was diagnosed with this disease, doctors told him he had two years to live; that was more than fifty years ago. He hadn't been out of his wheelchair since the late 1960s, but in 2007, when Peter suggested this flight to him, he leaped at it. Figuratively, of course.

Peter had created the Zero Gravity Corporation with his partners Byron Lichtenberg (who flew with my father on the shuttle) and Ray Cronise in 1994. The 727 flies in a parabola; it climbs at a 45-degree angle to about 32,000 feet, then throttles back. The pilot pushes the nose over slowly and the plane dives steeply, at a 30-degree angle, which results in about twenty-five seconds of total weightlessness. During those seconds passengers float as if they were in space, and the feeling is exactly the same for everyone, whether you are an Olympic athlete or a man who hadn't been out of a wheelchair in four decades. It's one of the most joyous experiences you can possibly have. For Stephen Hawking it was much more: the flight offered him twenty-five seconds of weightlessness after forty years of helplessness. It would be the first time in decades he could move as freely as anyone else. But there were also real dangers: anyone who has any trouble breathing, as Hawking does, experiences some difficulty at the bottom of the dive, during the pullout. We knew we could lighten the g-forces by planning a longer, shallower pullout, but until we actually flew, there was no way of knowing exactly how his body would react.

We met with him at the Kennedy Space Center for the first time two days before the flight. When you're speaking to him, it's impossible to tell what he's thinking because he's unable to make eye contact very well and he can't change his expression much. Obviously, he has no verbal or body language at all. So while you're sitting in front of him talking, he might appear to be looking at the ceiling behind you; you can't tell if that's simply a physical manifestation of his disease, part of his defensive nature, or the fact that he's not paying any attention to you. There is no real-time response, as he's not physically capable of that.

What would you say to Stephen Hawking when you met him? What would you talk about with him? I'm not easily intimidated. I grew up with American legends as my neighbors, but having a one-way conversation with one of smartest human beings on the planet was disconcerting. I kneeled down next to his wheelchair and introduced myself, then said, "Just like everybody else who's said they're a big admirer of yours, I'm a big admirer of yours." Then I did what I suspect everybody else does; I tried to fill in the silences. I did not know if I was actually communicating with him. I have no idea what I said to him. When I walked out of there, I felt like someone who had just finished an interview for the most important job of his career and had blown it completely. I couldn't stop second-guessing myself. Well, I thought, that was a once in a lifetime opportunity that I just messed up. I said exactly the same thing everybody else said. And then I thought, How incredibly boring that must be for him.

The day before Stephen's flight we made a rehearsal run at the Kennedy Space Center with a young boy of about his same size and mass. There were four emergency doctors and two nurses on that flight. The young volunteer was carried onto the plane acting as limp and unresponsive as possible. Stephen would not be able to have his special wheelchair or his communications gear on the flight. For the duration of the flight, he would be even more cut off than usual. We decided where to place him in a seat for takeoff and landing, how and where we would move him for zero-g parabolas, how to monitor his blood pressure and pulse, and, finally, how we were going to set up the cameras.

That night we had a little dinner in Stephen Hawking's honor. That was the first time I'd heard him "speak" in person. It was clear he heard everything being said, but his responses came about thirty minutes after events transpired. That was when I began to understand that Stephen really is a very witty guy—but on a thirty-minute delay.

He gives presentations by prerecording segments that he can queue up as he desires and then weave into paragraphs. That enables him to tailor his remarks to fit his audience. That night he "told" us about his life in general and specifically how proud he was to have the Sir Isaac Newton Chair at Cambridge. Then he began talking about his research. While this was a group of smart, very successful people, it was immediately obvious he had overestimated our technical acumen. He started explaining some of his most recent thinking about the nature of black holes—and as well-informed as I like to believe I am on this subject, he was going way over my head. I remember wondering, Does he actually think there's any possibility we know what he's talking about?

The next day Stephen was carried with great care aboard the plane. I took my position several feet from him. The flight took off and climbed into position for the first dive. We were all experienced Zero-G fliers and we formed a sort of interlocking spider's web, linked by our arms and legs—not to protect Stephen, but rather to ensure that each of us had a completely independent, unobstructed view of him floating freely in space. But we never forgot our primary objective: don't kill Stephen Hawking.

We all held our breath during that first parabola. The event could turn very quickly from something glorious to a disaster. We knew we could complete at least one parabola—twenty-five seconds of weightlessness was guaranteed—but there was no way of knowing what might happen after that. After the first period of weightlessness, we were going to fly straight while doctors evaluated him. If he felt capable of trying it a second time, we would do it. But our limit was three or four because we didn't want to push his boundaries. Peter Diamandis, Byron Lichtenberg, and several doctors stayed by Stephen to keep him in the middle of the plane and lift and turn him while the rest of us formed our web.

Everything went perfectly. Stephen lifted up, floated for a few precious seconds alone in the middle of the airspace, and gently settled back onto the mat with absolutely no difficulty. Well, it did require a complex maneuver to help him land in the position that was best for his breathing. But as soon as he was back down on the mat, it was obvious that this was going to be an even better experience than any of us had hoped for. Previously, I'd never seen even a slight change in his facial expression, but while he was in the air he had the biggest smile on his face; it was a huge smile and we all shared it. It was one of the most magical moments imaginable. He was free! For the first time in forty years, he was completely free. All my emotions kicked in: I felt overwhelming joy at being a part of this.

After we leveled off the doctors swarmed around him, testing his heart rate and blood pressure and trying to get a sense of his intention. Without access to his computer, it was difficult to communicate with him. Ultimately the doctors told us that he was absolutely fine, and one of his caregivers smiled and said, "Stephen wants to do more."

We did a second parabola. And a third. "Stephen wants to do more." It was as beautiful a time as I have had in my life. We did a fourth run, a fifth run. Um, Stephen, we should probably start wrapping this up. "Stephen wants to do more." A sixth and a seventh. We really need to wrap this up. An eighth. We were in the air for more than two hours and he was able to float, weightless, in space for a total of more than four minutes. His blood pressure never went up, his heart rate never went up or down; physically the experience did not seem to affect him at all. Someone had brought a few apples, and as a tribute to Isaac Newton and the man filling his chair, we let several of them float free in the cabin.

"It was amazing," Stephen said about thirty minutes after we had landed. "The zero-g part was wonderful and the higher-g part was no problem. I could have gone on and on. Space, here I come."

A year later, as I was preparing for my flight aboard the ISS, Space Adventures Vice President Tom Shelley told me that Stephen and his daughter Lucy were about to release a children's adventure book, *George's Secret Key to the Universe,* about a young boy who travels into space. Lucy actually wrote it, but at the end of each chapter, Stephen and several of the other best minds on earth wrote a paragraph explaining the real science that made the action in that chapter possible. The book is full of fascinating information about the universe. At that time I was putting together my list of things I intended to take into space and Tom suggested I bring Stephen's book.

Lucy and Stephen thought it was a capital idea and invited me to tea at their home. Most of my interaction that day was with Lucy, and although I was sitting right next to Stephen, Lucy did have to remind me to turn and speak directly to him. "Don't talk

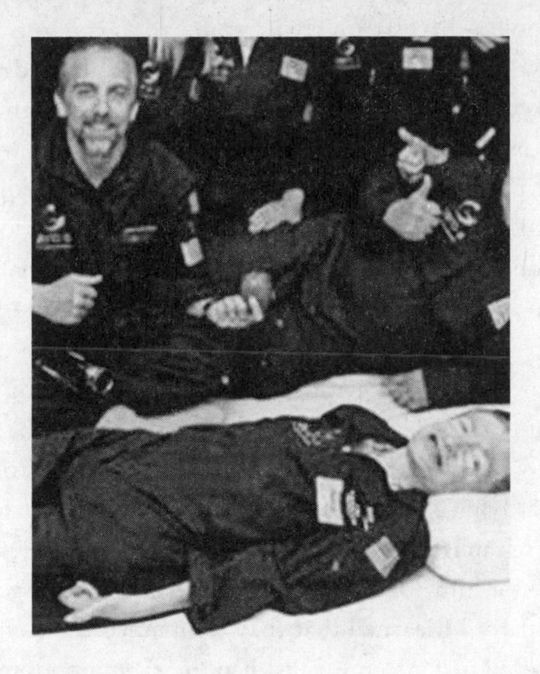

With Stephen Hawking on the day of his flight.

about Stephen," she told me. "Talk to Stephen. Don't worry, he'll process all of it later."

I spent most of an afternoon with them. I was a bit disappointed when I left that I hadn't been able to make a true connection with him on a personal level. Lucy walked outside with me and we were standing on the steps when an assistant brought Stephen out in his wheelchair. For the first time he said something specifically to me; I don't remember his exact words, but it was something like, "Richard, I thank you for coming all the way out here to see us. This is a wonderful idea. Thank you very much." It was one of those moments . . . I almost wanted to break down in tears right on the spot. To know that I had been able to make real contact with him, personal contact, was incredible.

The book itself was too big for me to carry into space—it would have taken up a sizable chunk of my allotted mass—so I just took the book jacket, complete with Lucy's signature and Stephen's thumbprint. After I got back I went to meet with them. This meeting took place in Stephen's office at Cambridge. Coincidentally, I had been born in Cambridge, only a few blocks away. Hanging on the door outside his office were several photographs of Stephen on his Zero-G flight. Then Lucy took me inside his office, where there are only a handful of things. One of them is a blackboard with his equations written on it in chalk. And as I looked at it I realized it had been shellacked so it couldn't be erased accidentally. This was the very last thing he was able to write on a chalkboard and it's been preserved. And right next to it were more photographs from our flight. "It was a completely profound experience for him," Lucy told me. "It's still something he thinks about regularly." I learned that they'd intended the cover to be a gift and had signed it to me, but I had the exact opposite idea and signed and gave it back to them, framed and with an ISS stamp.

What I do have are photographs and memories of that flight, and a desiccated apple. I had taken one of the Isaac Newton apples and put it in a refrigerator by my bedside. For more than a year I watched it shrivel, wondering all the time if it was going to rot, or turn moldy and force me to throw it away. Instead it freeze-dried; it shrank down to the size of a large walnut. So the question became, what to do with it? A specialty framer created a shadow-box frame for it, and it is hanging in my home. I've got this really cool framed apple in a box with a photograph of Stephen and me and it helps me retell the story of that flight with a little more passion.

And every time I look at it I can re-create the emotions of those moments in the airplane.

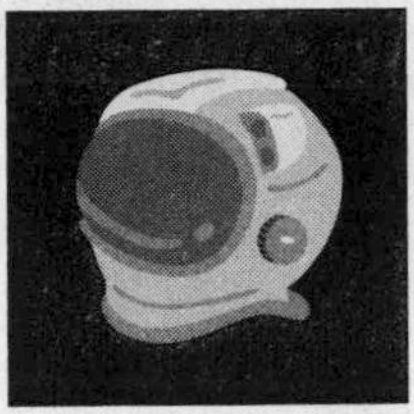

21 EXPLORE

Oh! The Places We'll Go

At some point in the future, I believe, people will look back at this period of discovery and exploration the way we look at the adventures of Marco Polo and Columbus.

We are at the dawn of a new age. As far as we have come, it is just the beginning. My two worlds, space and gaming, are both focused on exploring the possibilities. Some of these worlds are real, some are created, and some are in between. I can't predict the future—nobody can—but I am trying to help build it.

When my dream of going into space was temporarily derailed by a doctor, I became determined to do it myself. That's why I initially got involved in the privatization of space. There were only two space-faring nations on earth then, and if you had asked anyone what it would require to mount a private expedition into space, they would have told you it was impossible. They would have pointed out that NASA requires hundreds of thousands of employees and a multibillion-dollar budget to run America's space program. Indeed, the usual response to bringing it up was laughter.

But I've always believed that one day space travel will be led by private industry. That's been the history of most exploration. Columbus was sponsored by Spain, but by Darwin's day it was private industry mounting expeditions. In fact, throughout history both governments and private corporations have offered large prizes to encourage invention and exploration. For example, the Orteig Prize of $25,000 was offered in 1919 to the first person who flew nonstop between New York and Paris; after several other adventurers had died trying in the attempt, it was won in 1927 by Charles Lindbergh. Since that time there have been similar smaller prizes offered for feats like flying a hot-air balloon around the globe, but nothing has been created in the last half century to compare to the Orteig Prize. And then I met Peter Diamandis and heard his vision for the X-Prize.

I really do believe that my friends and I have fundamentally changed the future of human space exploration. In less than a decade the concept that private industry would make space travel possible went from being laughable to being everyone's prediction for the future. Things like asteroid mining and private telescopes have become the next possibility. The Google Lunar X-Prize will put more private vehicles on the moon to take up residence near my rover. There's no question that low-earth orbit and near-earth exploration have already turned from being government run to being almost the exclusive purview of private industry. The next question is what will happen with more ambitious expeditions.

Our next objective must be Mars. My good friend and the scientist/CEO I have the most respect for on this planet, Elon Musk, already has rockets flying cargo to the ISS, and is building a vehicle called the *Red Dragon* to take humans there. The Russians are planning a trip as well, although budget issues have hampered them. So odds are that the first trip outside the earth's gravity is going to be accomplished by private industry.

How will NASA fit into this? My own belief is that it should be the responsibility of the government to open new territories. The main reason for the government to go somewhere is to discover if there is potential value there. We have been to the moon, we know what is there, and industry can either take advantage of what's there or they can't. But the government doesn't have to continue going back. If there is value it should be up to private industry to exploit it, because that allows the government to continue to pursue more speculative goals. Those are the kinds of risks I believe government should be taking.

Living on Mars would quickly expose a human to their lifetime radiation limit. Beyond that are the giant gas planets, and it's not clear why we would want to go there. But there are some amazing moons around Jupiter and Saturn that might be interesting to explore. If we are looking for volunteers to go on one of these voyages, I'm raising my hand. If I could be one of the first colonizers on Mars, I would do it. And in the best of all possible scenarios, my family would go with me.

The Garriotts' Chariots

When my wife, Laetitia, and I first began dating, she made it clear to me that she had no interest in going into space. No way, no how. As her grandfather was the first planetary geologist—the Cayeux crater on the moon is named after him—one might believe she would obviously have an interest in going. And one would be wrong. She made it clear to me that she loved life on earth far too much to leave it behind or take the risks inherent in space exploration.

But then she added that there was also no way that I was going back into space. As a father, she reminded me, I have obligations here on earth that take precedence for now.

I suspect we are probably one of the first couples to argue seriously about whether or not we would go into space. I do want to go back one day, and I suspect she may change her mind about it.

She has already come a long way. In 2011, we went to Russia

for a series of official celebrations surrounding the fiftieth anniversary of Yuri Gagarin's historic flight. By the time we were ready to return to the U.S., Laetitia had become the youngest person to complete neutral buoyancy training in Star City. She spent over an hour in this deep pool with a full-size mock-up of the space station, in full cosmonaut gear, executing with determination and great poise every task the control center threw at her. She was so impressive the Russians suggested that if the European Space Agency selection team had been there, she would be preparing to board a flight.

Laetitia smiled and thanked the Russian team, but she reminded me that she still loved life on earth way too much.

But for the first time she understood my passion, and had begun to feel it herself. And I had now shared with her a practical side of the joy of space exploration—one that she rates to this day as among one of her pinnacle life experiences—and that was very special to me.

Given her family background it isn't surprising that Laetitia has always been intrigued by the mysteries of space. But if she wasn't going to travel in space, as other women have done, she still intended to have an impact. As I learned a long time ago, if access to space is to become readily available, significant changes have to be made in safety and cost. And pioneering those changes is the path she has taken.

Russian scientist Konstantin Tsiolkovsky is famed for "the rocket equation," the essential theory that has formed the foundation of the first half century of space exploration. But in *The Spaceship,* published in 1924, he proposed that the most efficient manner of going into space would not be to use chemical propulsion, as we do now, but rather to beam energy—electromagnetic rays of short wavelength—to a spacecraft.

It's the stuff of science fiction—and working with a team of Caltech scientists on it is what Laetitia has set out to do. The company she cofounded, Escape Dynamics, has proved it is possible, building a prototype thruster that operated when energy from the electric grid was converted into a microwave beam and beamed to a thermal thruster, which generated a highly efficient thrust. The feat was not small. Their successful experiments made the cover of *Aviation Week* magazine. *Scientific American* described their beamed-energy propulsion concept as a "World Changing Idea" that could enable microwave-powered space planes and aircraft-like operations to reach orbit. Although the cost of completing the research and design made the concept economically impossible, causing the company to close, Escape Dynamics had proved that microwave propulsion not only is feasible, but in fact is capable of efficiency and performance surpassing traditional chemical rockets. That is a huge step into the future—and in response NASA has added beamed-energy propulsion to their roadmap.

Laetitia has actually become more of an expert than I am. And if this work did not change her personal comfort zone about going into space, it clearly was opening a new path into space for me and many others.

Whether or not I manage to go again, or whether Laetitia goes with me, I am confident our children will travel in space. If I have to, I'll lock them in a tin can and shoot it up myself.

Where in the Universe?

I enjoy thinking about permanence. Under my mother's tutelage, three decades ago I made a necklace and a chain and put it on. I still haven't taken it off. Part of the reason for that, of course, is that I didn't know how to make a clasp—but I suspect I would have left it on permanently anyway. So for me it's fun to speculate that someday my eighth-generation descendants will note while looking back at earth from wherever humanity has spread by then, that yes, their great-great-great-great-great-great-great-great-grandfather Owen Garriott was the 63rd person to go into space and that I, his son, was the 483rd person to travel in space, and therefore our family of spacefarers helped build the foundation for our space society.

My descendants are going; the only question is where. The sensible assumption is that we will go wherever we need to go in pursuit of life in any form. It is almost impossible for me to believe that there isn't life on other planets. I've spent considerable time recently speculating about the possibility that at one time life existed on Mars. It's the only other planet in our solar system that is in the Goldilocks zone, meaning it once had an atmosphere and liquid water. It had a magnetic field that offered some protection. And there is more and more data being reported that allows nonexperts like me to imagine how or why life may have existed on Mars.

Based on the evidence, I have answered that question about life on Mars to my own satisfaction—but that answer seems to change based on the day you ask me, or even the time of day. For a long time I believed that because there had once been water on Mars, it was reasonable to assume that where there is water there is life. And Mars formed about the same time earth did, so if life is common in our universe, it is likely it would have been there too. "Life" in this instance does not mean little green men but rather trace amounts of microbes. While visiting the thermal vents on the ocean floor with my father, we discovered unknown life forms capable of surviving in that extraordinary heat and pressure, so I at least accept the possibility that life forms could survive on Mars.

I have spent my life being fascinated by the history of life on earth: Life began when two planetary bodies collided, or a small planetoid hit the early forming earth, and spewed out a bunch of the earth's crust, which became our moon. The earth at that time was completely molten, so if life had existed before that—which is doubtful—it was eradicated. But when the earth and moon began to cool, as soon as there was a crust on earth and rains began to fall and create the oceans, life effectively began.

That surprised me. If life is that springy, if it will manifest that quickly, it's very likely that there was life on Mars. There was water on Mars for long enough. So I began gathering additional data. Mars is only a third of the size of the earth, and it turns out that Mars lost its magnetic field, its atmosphere, and its liquid surface very quickly, in just a couple of hundred million years. There has been life on earth for over 4.6 billion years, so if Mars had a liquid surface for only 200 million years, I'm not sure that was enough time for life to appear. All of a sudden I wasn't so confident.

I began trying to find out how long it was after our oceans came into existence that life on earth began. It turned out to be immeasurably close to the beginning, within 100 million years. And the length of time that Mars had a liquid surface would just fit into that span of time. So maybe there was time. Every little bit of data flipped me back and forth. One thing is for certain, we are going to Mars, and when we get there, we'll be able to make a pretty good determination as to whether or not life existed. Meanwhile, we are searching the skies, hunting for evidence of some form of life on other planetary bodies. To paraphrase Dr. Seuss, Oh, the places we'll go.

Entertaining New Ideas

Here on earth I expect to continue creating new worlds of entertainment. I believe the gaming world, like space travel, is just in its infancy. The promise as well as the potential of this art form goes well beyond players shooting bad guys. A major criticism of computer games is that they encourage young people to sit alone in a dark room instead of going outside and doing something active with friends. And I think that criticism is fair when applied to early video games. The first two decades of this amazing digital age were very much like the early years of television, when people migrated from being outside doing whatever they were doing to sitting passively inside in front of a box. While video games are a little less passive than watching TV, that criticism is probably valid; people do need to interact with other real people. But I also believe that online games are now reversing that trend in a really important way.

Throughout history we have lived in tribes of several hundred people who banded together in some form of community to survive. As societies grew, of necessity they formed new tribes. People lived close to their jobs, and those jobs often were in the same or related industries. So mill towns and coal towns developed, as well as the service businesses needed to provide support. People were brought together by sharing a common experience, and people got to know their neighbors, who were often part of the same social group.

That changed drastically with the creation of a commuter society, as cars and mass transit expanded the distance people could travel easily and quickly. People had much less in common with their neighbors than they once did. The commuting lifestyle most Americans led geographically separated people with common interests. What computer networks do, what social networks do, is take people who have found themselves displaced from their tribe and give them a way to get back together. They have made it possible for them to find their tribe online.

While it is not unreasonable to describe early offline games as working against social bonding, online communities have now become a unique and effective means of forming social coalitions. MMORPGs in particular have brought people with common interests together. That's going to continue, but what will evolve is the environment in which these relationships are fostered and played out. Way, way in the future there will be the *Matrix*-style plug in, and slightly sooner there will be a simple, transparent version of the holodeck of the *Star Trek* experience, while in the very near future we'll get to play with goggles and data gloves. As a creator of virtual experiences, I am fascinated by what this path holds.

Anyone working in the gaming industry is also in the business of predicting the near future. Because of the time it takes to make a game, we have to predict which platforms will be popular. That means we need to know what hardware is going to be readily available three years from the day we start making a game; and when you're talking about technology, three years is an eternity. So in our industry we make very high-stakes stabs in the dark.

I do have some sense of where we'll be in the distant future. The next long-term step is the continued development of virtual reality technology. Early in my career I was very optimistic about this coming soon. I thought people would be able to sit in a room and, through this amazing technology, believe they were in a projected world and interact with it. A joystick or data gloves would allow them to have contact with virtual items in that space. So for many years I bought every virtual reality device that came out, hoping we were getting closer to that day. But, unfortunately, we still haven't reached the first stage of virtual reality. The level at which my senses are completely fooled, and I forget that I'm not in the real world, is still far away.

Imagine the possibilities. Having a version of the *Enterprise*'s holodeck in your playroom, for example. A system that would enable you to play a perfect game of tennis—against Björn Borg. Or bat against Sandy Koufax. Or explore the surface of Mars. Or that would let me take you through my haunted house without actually having to build the house. The experiences that we will be capable of programming will enhance and enrich lives in ways that are not yet unimaginable.

Obviously the storytelling possibilities are going to be amazing. What started as paintings on the wall of a cave have become the color of history; we describe entire civilizations through the few of their stories that have been passed down to us. And new technology will enhance the way we will tell the great stories of our civilizations, both real and imagined. We'll be able to put players inside the games that I have been creating for my entire career: Britannia really will be a place you can go. Most exciting to me is that while the technology will evolve, well-told stories will always be the foundation of the experience.

These changes are already happening incrementally. Until recently, for example, people couldn't play MMOGs unless they were online—that's what that *O* represents. When I was on an airplane, for example, I couldn't play my own game. But I wanted to be able to play anywhere, at any time. A lot of people like to play solo anyway; they frankly don't give a flip about who else is online, they themselves just want to play. And like many people, I find myself playing games on mobile devices more and more. So when we were developing our latest game, *Shroud of the Avatar,* we incorporated technology that allows people to play offline; as soon as they go online it updates the permanent changes to the cloud. In addition, if two people want to play together, they can directly connect with each other instead of meeting in the cloud. But while there may be only two people playing together, they will still see what others are doing. If somebody builds a new shop that sells great swords, for example, those two players will be able to walk inside and buy a sword. This is a minor technological advance, a step in the right direction.

One evening at the Explorers Club in New York, I was sitting with the late astronaut Scott Carpenter, the second man to orbit the earth, and I said to him, not at all facetiously, "I know what you're known for, but I want to know what other contributions you've made toward advancing the space program, other than being the second guy."

It was obvious he liked the question. "I'll tell you what I believe my contribution was," he said. "Of all the original astronauts I was the hardest-core engineer. We had a bunch of young engineers building these rockets all of us would fly in the *Mercury* program. While a lot of us were test pilots, I was the strongest engineer. And so I was the person who was always digging through the output of those bright young engineers and going to the furthest detail of perfection. We were all putting our lives on the line when we climbed aboard these machines, we all knew there was a lot of inherent risk, so what we did not want to do was compound that inherent risk with engineering mistakes." Scott Carpenter was the strongest proponent of all the checks and balances that ensured the engineering discipline was as tight and strongly managed as humanly possible.

I have thought long and hard about my own contributions in the past and how to continue to make contributions in the future. Clearly my contribution to space travel has nothing to do with the fact that I actually got to go, it's what I did in order to be able to go. It's what Laetitia is continuing to do. It is likely that in the distant future no one will be playing my games, but the concepts and features that I have championed have already made a lasting impact.

Here's what I know: I want to continue doing what I have been doing, which is deriving great joy from creating experiences. I want to continue exploring the earth and space in search of the wonders that await us all. And I want you to spend your life in pursuit of what fascinates you. It will be there . . . the question is, will you notice, will you stop and take a look?

Acknowledgments

Thank you to:

My amazing coauthor, David Fisher, without whom you would not be reading this book today.

I am a mere storyteller; this book wouldn't be what it is today without the penmanship of David with whom I have enjoyed long conversations over the past few years.

I also want to acknowledge:

My mother, Helen, who inspired my deep curiosity and artistic expression.

My father, Owen, who helped me tackle the tough science and math problems and split the cost on my first computer.

My brother Robert, for suggesting we start Origin, which grew to one of the great development houses of the early games industry.

My sister, Linda, who was my best buddy growing up, for tolerating and surviving the dangers and scrapes I dragged her into.

My youth theater coach, Claire Harmon of Clear Creek Country Theater, who helped craft my confident extroversion.

Organizations like the Boy Scouts, Camp Manison, and Junior Achievement, which played a big part in preparing me for my future. And, of course, all of my friends at the Society for Creative

Anachronism for borrowing their names and life stories for inspiration as I created the characters in my games.

The teachers, parents, and friends who played D&D with me every Friday and Saturday for years as I crafted my first worlds.

My amazing and talented game aficionados—programmers, artists, designers, and other developers, as well as my space cadet partners in exploration companies—with whom I have shared this journey.

I know well that little of this journey would exist without you all.

Creating with you all has and remains one of my life's greatest joys.

The players of my games and readers of this work.

If we have not yet met, I hope we do.

Those who know me know that I celebrate the bonds of community we now share, whenever I can.

When you finish this book—as I have said with all my games—let me know. I will do my best to respond in kind.

And I can't wait to hear where you'll go and what you'll build!

David Fisher Would Like to Acknowledge

Each project allows me entrance to new worlds, but never have I had the privilege of exploring so many new worlds at one time as I have in working on this book. It remains a true delight to have worked with and gotten to know Richard and Laetitia Garriott de Cayeux, extraordinary people whose many kindnesses—and continuing enthusiasm—I greatly appreciate. Visiting the many places in Richard's mind and through his adventures has been a great pleasure and I want to thank Richard for having me as a guest there.

In my travels through Richard's world I have had the enjoyment of speaking to many of his friends and the people with whom he has worked. I would like to especially thank Robert Garriott, Starr Long, and Dale Flatt for their time.

I also want to thank Casson Masters and Scribecorps for assistance in transcribing the many interviews with extraordinary accuracy. I also appreciate the editorial assistance of Beau Stevens in getting all of this done.

This project was initiated by the fine literary agent Jeff Silberman, who was relentless is turning his vision into this book, and his efforts are valued.

We also have been very fortunate to have worked with Peter Hubbard, executive editor at William Morrow, as well as Nick Amphlett, whose editorial expertise is visible on every page. Richard and I both greatly appreciate Nick's efforts.

In addition, I want to acknowledge the contribution to my peace of mind in this and other projects made by Alan Susskind of CSD Associates. Knowing he is there allows me to focus on those things that need to be done. His work and his decades-long friendship is greatly valued.

And finally, always last but never least, is my wife, Laura. Nothing I do, nothing I accomplish, is possible without her. I appreciate her unfailing support while creating her own career. I want her to know I am extraordinarily proud to be her husband.

When Richard and I first met we found we shared a great love and appreciation for creative vision, and we wanted to at least tickle that in readers' minds. We both have great hope we have succeeded in doing that.

About the Authors

Richard Garriott founded the gaming companies Origin Systems, Destination Games, and most recently Portalarium. His *Ultima* series of games has sold millions of copies worldwide. In 2008, Richard rode a Soyuz TMA-13 to the International Space Station, becoming the first American second-generation space traveler. He lives in New York City with his wife, Laetitia, and their two children.

David Fisher is the author or coauthor of dozens of books, including twenty-two *New York Times* bestsellers. He has worked with George Burns, Johnnie Cochran, and Terry Bradshaw, among others. He lives in New York City.